ADDITION

Applying addition strategies

Published by World Teachers Press®

Printed in the United States of America.

This book is printed on recycled paper.

Order Number 2-5239
ISBN 978-1-58324-203-2

B C D E F 14 13 12 11 10

395 Main Street
Rowley, MA 01969
www.didax.com

Foreword

Addition is a compilation of activities designed to introduce addition and provide strategies for numbers to 20 and beyond. Initial activities encourage the use of concrete materials to establish the concept of addition through "real life" number stories. The activities which follow allow students to consolidate the addition concept while providing a pool of strategies to promote mental and written computation.

The simple format of each task allows students to work independently or in small groups. Consolidation tasks have been included to allow students to practice strategies they have learned mentally and in an abstract written form.

Assessment pages have been included to assist in the monitoring of student progress. These activities are also suitable for inclusion in student portfolios. The activities included in *Addition* will complement the teaching and learning of addition in the lower elementary classroom and help students create a solid foundation on which to build further numeration skills.

Titles in the series:

Addition *Grades 1 - 2*
Subtraction *Grades 1 - 2*

Contents

Teacher's Notes 4
Checklist – Addition Concepts and Strategies 5
Number Stories for 2 ..... 6 - 7
Number Stories for 3 ..... 8 - 9
Number Stories for 4 ..... 10 - 11
Number Stories for 5 ..... 12 - 13
Number Stories for 6 ..... 14 - 15
Number Stories for 7 ..... 16 - 17
Number Stories for 8 ..... 18 - 19
Number Stories for 9 ..... 20- 21
Number Stories for 10 ..... 22 - 23
Assessment – Number Stories 24
Count up 1 ..... 25 – 26
Count up 2 ..... 27 – 28
Count up 3 ..... 29 – 30
Equal 31
Adding Zero 32
Review – Count on Race 33
Review – You're Joking 34
Swap Around Facts – Count on 1 35
Swap Around Facts – Count on 2 36
Swap Around Facts – Count on 3 37
Swap Around Street 38
Doubles 1, 2, 3 ..... 39 – 40
Doubles 4, 5, 6 ..... 41 – 42
Doubles 7, 8, 9 ..... 43 – 44
Double Trouble 45
Review – Doubles ..... 46 – 47
Double and Count up 1 48
Doubles and Doubles Count up 1 49
"Magic 10" Facts ..... 50 – 51
Using "Magic 10" Facts 52
Building to 10 53
Adding onto 10 54
All the Facts to 10 ..... 55 – 56
Speed Test – All the Facts to 10 57
All the Facts to 20 ..... 58 – 59
Speed Test – All the Facts to 20 60
Strategies for Adding to 10 61
Strategies for Adding to 20 62
Answers ..... 63 – 64

Teacher's Notes

Addition has been sequenced to ensure a smooth transition for young learners, allowing them to develop a conceptual awareness of adding and then apply this to mental and abstract tasks.

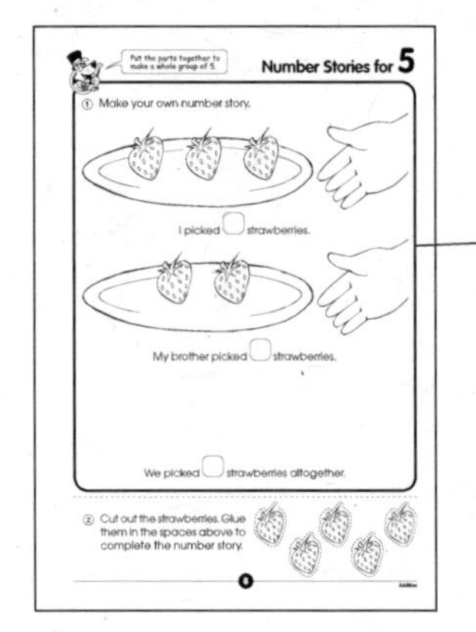
Number Stories for 5

1. Make your own number story.

I picked ☐ strawberries.

My brother picked ☐ strawberries.

We picked ☐ strawberries altogether.

2. Cut out the strawberries. Glue them in the spaces above to complete the number story.

The addition concept is introduced using pictorial aids to support number stories. These activities require students to experiment with number combinations as parts being joined to make a whole. To enhance these activities, concrete materials should be provided to allow students tactile and visual stimulus for consolidating their knowledge of the "part-part-whole" relationship.

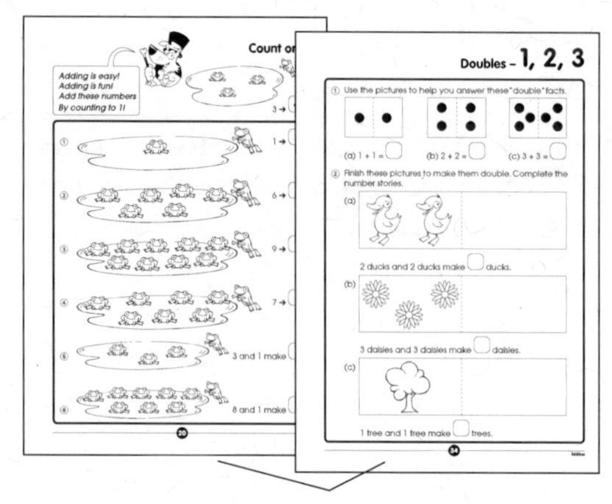
Count on

Adding is easy! Adding is fun! Add these numbers By counting to 1!

3 and 1 make ☐

8 and 1 make ☐

Doubles - 1, 2, 3

1. Use the pictures to help you answer these "double" facts.

(a) 1 + 1 = ☐ (b) 2 + 2 = ☐ (c) 3 + 3 = ☐

2. Finish these pictures to make them double. Complete the number stories.

(a) 2 ducks and 2 ducks make ☐ ducks.

(b) 3 daisies and 3 daisies make ☐ daisies.

(c) 1 tree and 1 tree make ☐ trees.

Consolidation pages support the learning of new strategies and allow students to practice the new skills they have learned in an informal situation. Number facts to 20 are regularly explored throughout these activities, encouraging the students to automatically recognize common number facts and, in turn, increase their confidence in addressing numeration tasks

A series of fundamental addition strategies, including counting up, doubling, and adding to 10, have been sequenced to maximize the student's confidence in adding and ability to add simple number facts to 20 and beyond. These strategies enable students to gradually begin building a repertoire of known number facts and provide a foundation for introducing other operations and numeration concepts.

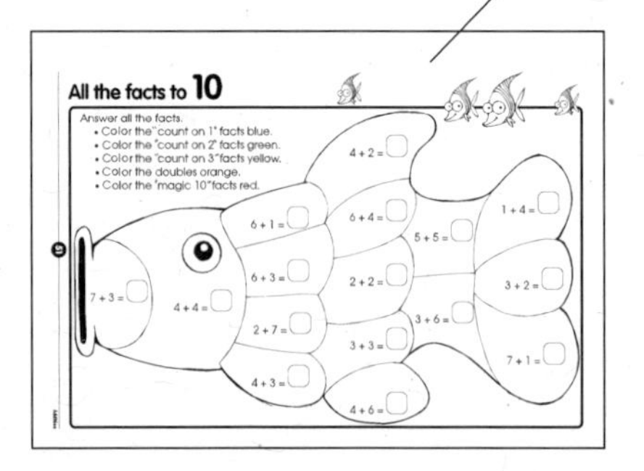
All the facts to 10

Answer all the facts.
- Color the "count on 1" facts blue.
- Color the "count on 2" facts green.
- Color the "count on 3" facts yellow.
- Color the doubles orange.
- Color the "magic 10" facts red.

4 + 2 = ☐ 6 + 1 = ☐ 6 + 4 = ☐ 1 + 4 = ☐ 5 + 5 = ☐ 6 + 3 = ☐ 2 + 2 = ☐ 3 + 2 = ☐ 7 + 3 = ☐ 4 + 4 = ☐ 2 + 7 = ☐ 3 + 6 = ☐ 3 + 3 = ☐ 7 + 1 = ☐ 4 + 3 = ☐ 4 + 6 = ☐

Assessment pages have been provided to enable you to evaluate the progress of individual students. These assessment pages provide a comprehensive summary of the strategies introduced in the book and are appropriate for use in student portfolios. A class checklist has also been included on page 5 to assist you in monitoring student understanding of the strategies taught.

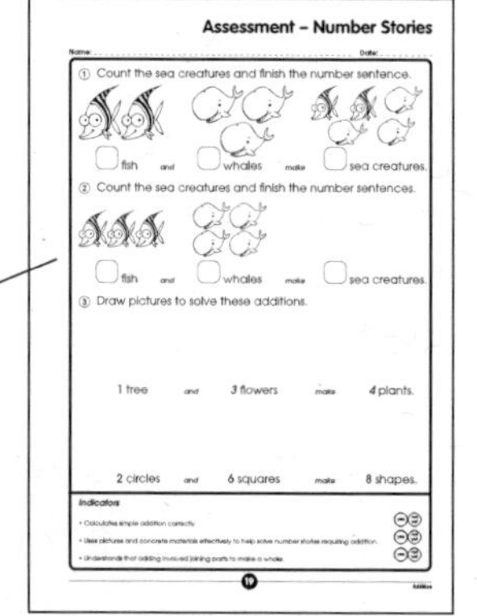
Assessment – Number Stories

1. Count the sea creatures and finish the number sentence.

☐ fish and ☐ whales make ☐ sea creatures.

2. Count the sea creatures and finish the number sentences.

☐ fish and ☐ whales make ☐ sea creatures.

3. Draw pictures to solve these additions.

1 tree and 3 flowers make 4 plants.

2 circles and 6 squares make 8 shapes.

Indicators

Simple rhymes have been included as an auditory reminder for students as they apply new concepts and strategies. These rhymes can also be enlarged and displayed in the classroom as a visual clue for solving mental addition facts.

To enhance the teaching and learning of new numeration concepts, you can model them for the students using concrete materials and ample opportunity for the students to practice and apply the concept.

Using a foundation of concrete experience when introducing each new concept will allow for a smooth transition into pictorially presented written tasks and subsequent symbolic and abstract representations.

Many activities encourage students to look for numerical patterns in each new strategy. Patterns found in mathematics tasks enable students to see and comprehend solutions to increasingly difficult tasks and form the basis of all mathematical strategies.

The introduction to the concept of addition on pages 6 to 23 is followed by more complicated strategies for older or more capable students who have already mastered the addition concept. These strategies may also provide a different understanding of the concept.

Checklist — Addition Concepts and Strategies

Student Names	Completes number stories	Uses concrete materials	Calculates mentally	Calculates abstractly	Understands P/P/W	Understands zero	Counts back 1	Counts back 2	Counts back 3	Swaps facts around	Understands halving	Subtracts 10	Knows "magic 10" facts	Finds "missing" parts	Compares groups

Number Stories for 2

Put the parts together to make a whole group of 2.

① Count the flowers and draw the total number.

1 rose	*and*	1 daisy	*make*	2 flowers.
part		part		whole

② Draw the insects and finish the number story.

1 bee	*and*	1 spider	*make*	☐ insects.
part		part		whole

Put the parts together to make a whole group of 2.

Number Stories for 2

① Complete the number story.

There is ☐ turtle in the pond.

Along comes ☐ more.

Now there are ☐ turtles in the pond.

② Cut out the turtles and glue them in the pond.

Number Stories for 3

Put the parts together to make a whole group of 3.

① Count the number of fish and draw the total in the tank.

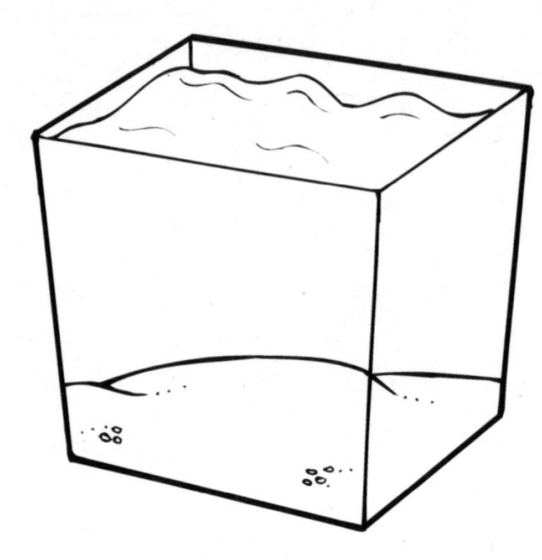

2 fat fish *and* **1** thin fish *make* **3** fish in the tank.

part — part — whole

② Draw the ducks and finish the number story.

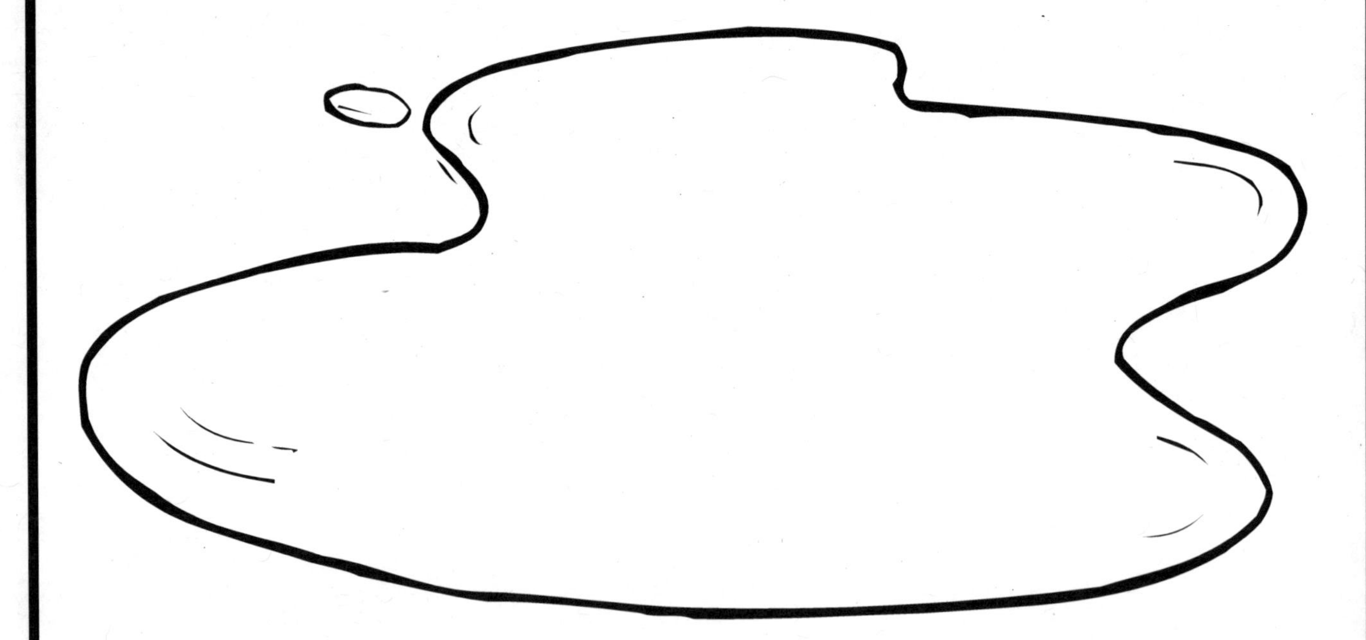

1 yellow duck *and* **2** black ducks *make* ☐ ducks.

part — part — whole

Put the parts together to make a whole group of 3.

Number Stories for 3

① Complete your own number story.

There were ☐ spiders in a web.

Along came ☐ more spider.

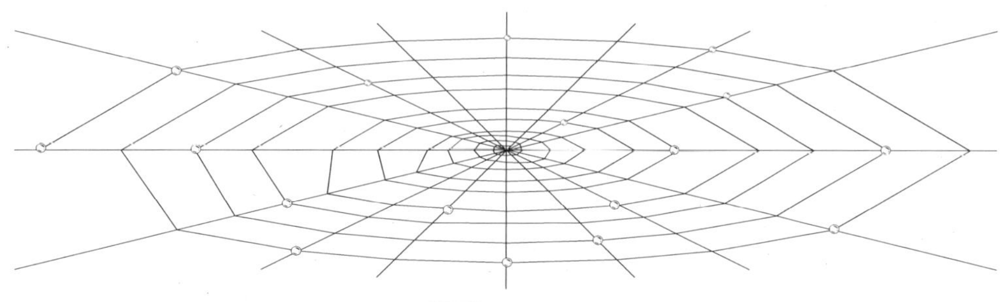

Then there were ☐ spiders in the web.

② Cut out the spiders. Glue them in to the web above to complete the number story.

Number Stories for 4

Put the parts together to make a whole group of 4.

① Draw the leaves on the plant.

1 leaf	*and*	3 leaves	*make*	4 leaves on the plant.
part		part		whole

② Draw the cars and finish the number story.

2 red cars	*and*	2 blue cars	*make*	☐ cars.
part		part		whole

Put the parts together to make a whole group of 4.

Number Stories for 4

① Make your own number story.

There were ☐ birds on a branch.

Along came ☐ more birds.

Then there were ☐ birds on a branch.

② Cut out the birds. Glue them onto the branch above to complete the number story.

Number Stories for 5

*Put the **parts** together to make a **whole** group of 5.*

① Count the fruit and draw the total.

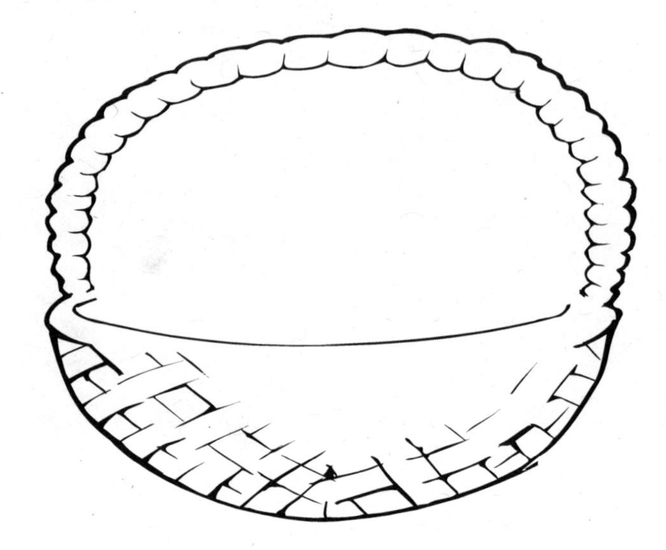

4 bananas *and* **1** apple *make* **5** pieces of fruit.
part · part · whole

② Draw the lollipops and finish the number story.

3 orange lollipops *and* **2** purple lollipops *make*  lollipops.
part · part · whole

Number Stories for 5

① Make your own number story.

I picked ☐ strawberries.

My brother picked ☐ strawberries.

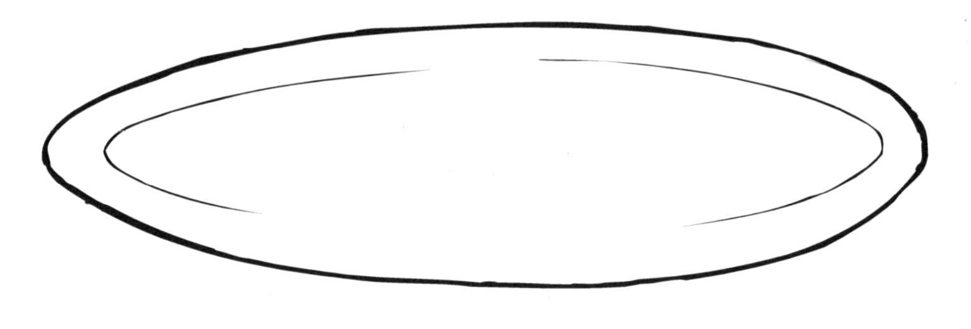

We picked ☐ strawberries altogether.

② Cut out the strawberries. Glue them on the plate above to complete the number story.

Number Stories for 

*Put the **parts** together to make a **whole** group of 6.*

① Count the trees and draw the total.

4 trees *and* 2 trees *make* 6 trees in a forest.
part — part — whole

② Draw the beads and finish the number story.

5 green beads *and* 1 blue bead *make* ☐ beads.
part — part — whole

Put the parts together to make a whole group of 6.

Number Stories for 6

① Make your own number story.

Chad collected ☐ beetles in the garden.

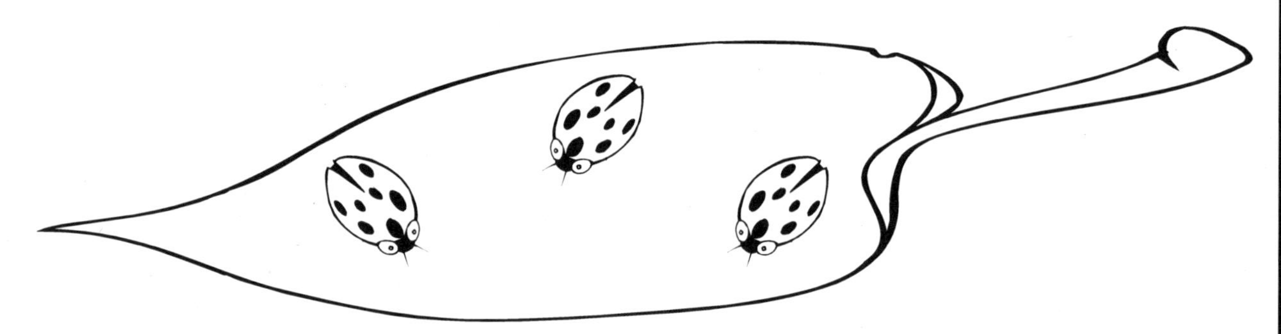

Ricky collected ☐ beetles.

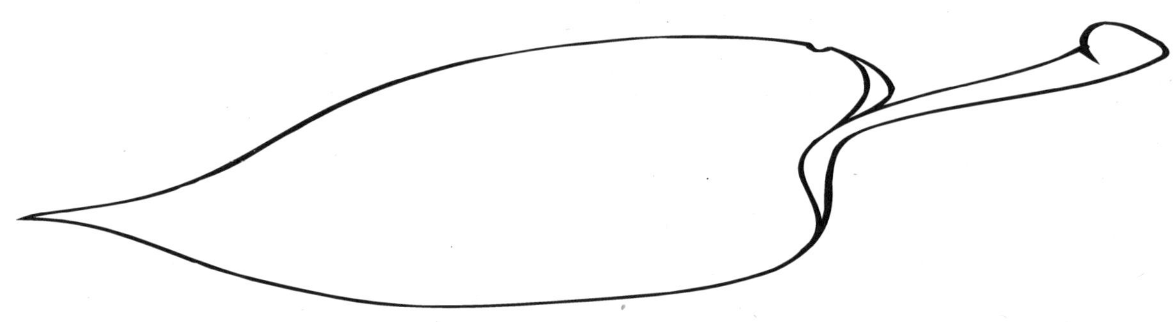

They put the beetles on a leaf.

There were ☐ beetles altogether.

② Cut out the beetles. Glue them on the leaf above to complete the number story.

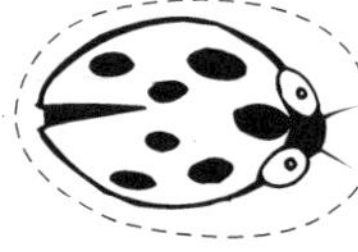
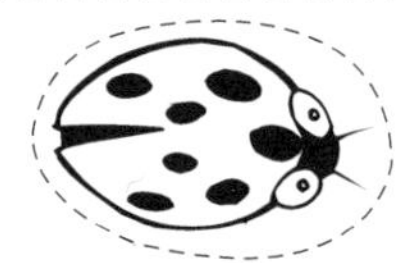
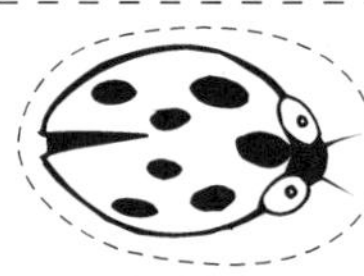
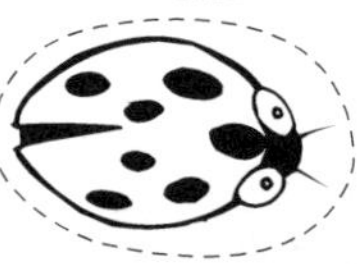
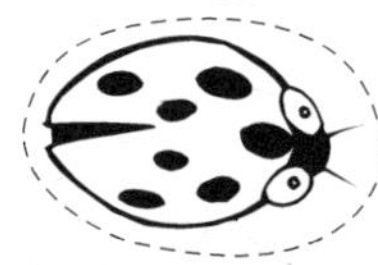
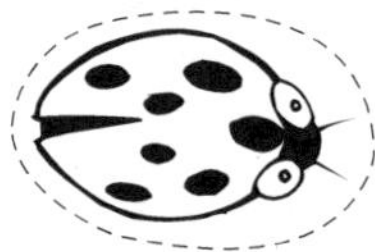

Number Stories for 7

*Put the **parts** together to make a **whole** group of 7.*

① Draw the children and draw the total.

3 girls *and* 4 boys *make* 7 children playing together.
part part whole

② Draw the vegetables.

5 carrots *and* 2 beans *make* ☐ vegetables.
part part whole

Number Stories for 7

① Make your own number story.

Jenny found ☐ lemons under the tree.

Annie found ☐ lemons.

They put the lemons in a bowl. There were ☐ lemons altogether.

② Cut out the lemons. Glue them in the bowl above to complete the number story.

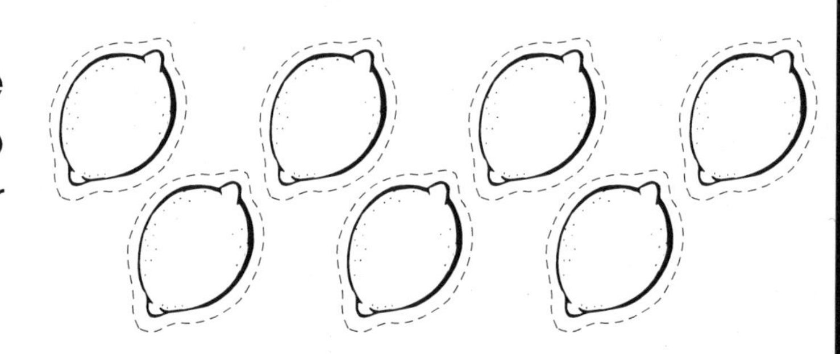

Put the parts together to make a whole group of 8.

Number Stories for 8

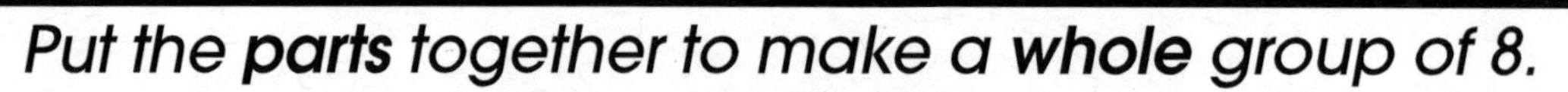

*Put the **parts** together to make a **whole** group of 8.*

① Count the stars and draw the total.

5 stars (part) *and* **3** stars (part) *make* **8** stars in the sky. (whole)

② Draw the insects and finish the number story.

2 black ants (part) *and* **6** red ants (part) *make* ☐ ants. (whole)

Put the parts together to make a whole group of 8.

Number Stories for 8

① Complete your own number story.

There were ☐ chickens in the nest.

Along came ☐ more chickens to the nest.

Then there were ☐ chickens in the nest altogether.

② Cut out the chickens. Glue them in the nest above to complete the number story.

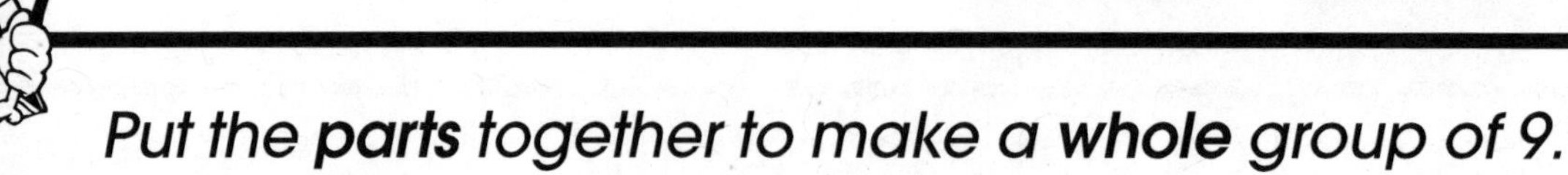

Number Stories for 9

Put the parts together to make a whole group of 9.

① Count the books and draw the total.

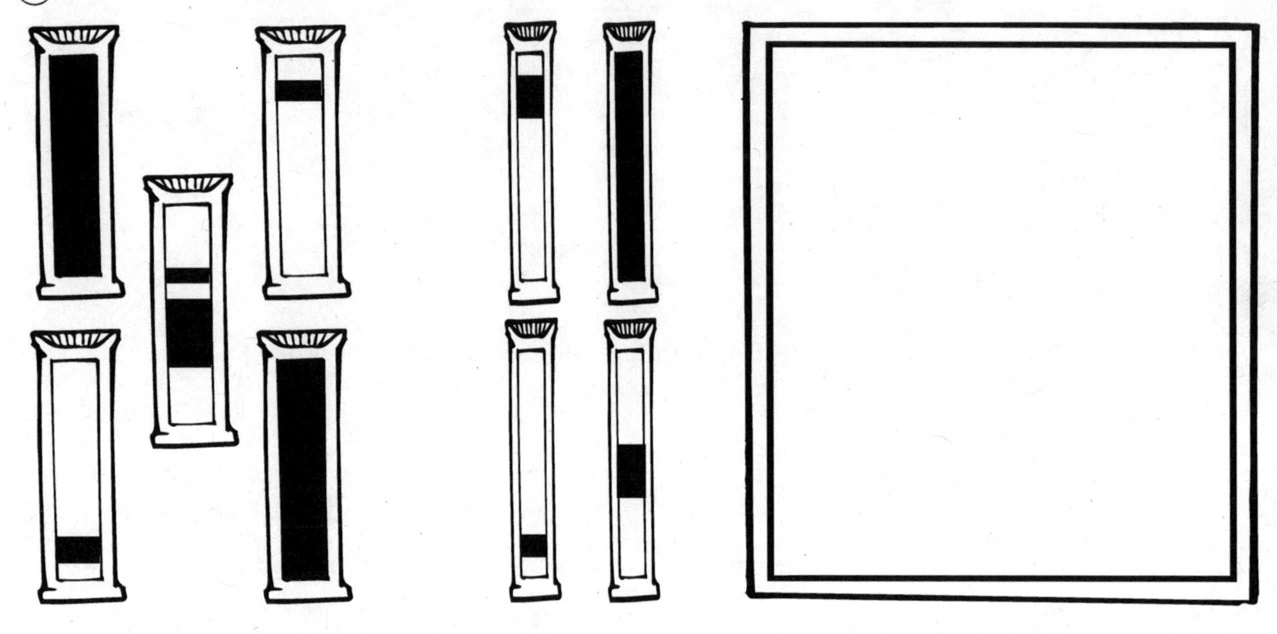

5 thick books *and* **4** thin books *make* **9** books on the shelf.

part — part — whole

② Draw the shapes and finish the number story.

6 circles *and* **3** squares *make* ☐ shapes.

part — part — whole

Put the parts together to make a whole group of 9.

Number Stories for 9

① Make your own number story.

Mom put ☐ cans of fruit in the cart.

I put ☐ cans of baked beans in the cart.

We had ☐ cans in the cart altogether.

② Cut out the cans. Glue them in the cart above to complete the number story.

*Put the **parts** together to make a **whole** group of 10.*

① Count the spots and draw the total on the shirt.

6 small black spots *and* **4** big black spots *make* **10** spots on the shirt.

part part whole

② Draw the eggs and finish the number story.

3 blue eggs *and* **7** yellow eggs *make* ☐ eggs.

part part whole

Put the parts together to make a whole group of 10.

Number Stories for 10

① Make your own number story.

Lewis found ☐ termites in the log.

Luke found ☐ termites in an old fence.

They put them in a jar. They found ☐ termites altogether.

② Cut out the termites. Glue them in the jar above to make up your own number story.

Assessment – Number Stories

Name: _______________________________ Date: __________

① Count the sea creatures and finish the number sentence.

☐ fish *and* ☐ whales *make* ☐ sea creatures.

② Count the sea creatures, draw the total and finish the number sentence.

☐ fish *and* ☐ whales *make* ☐ sea creatures.

③ Draw pictures to solve these addition problems.

(a) **1** tree *and* **3** flowers *make* ☐ plants.

(b) **2** circles *and* **6** squares *make* ☐ shapes.

Indicators

- Calculates simple addition correctly (yes / not yet)
- Uses pictures and concrete materials effectively to help solve number stories requiring addition. (yes / not yet)
- Understands that adding involves joining parts to make a whole (yes / not yet)

Count up 1

Adding is easy!
Adding is fun!
Add these numbers
By counting up 1!

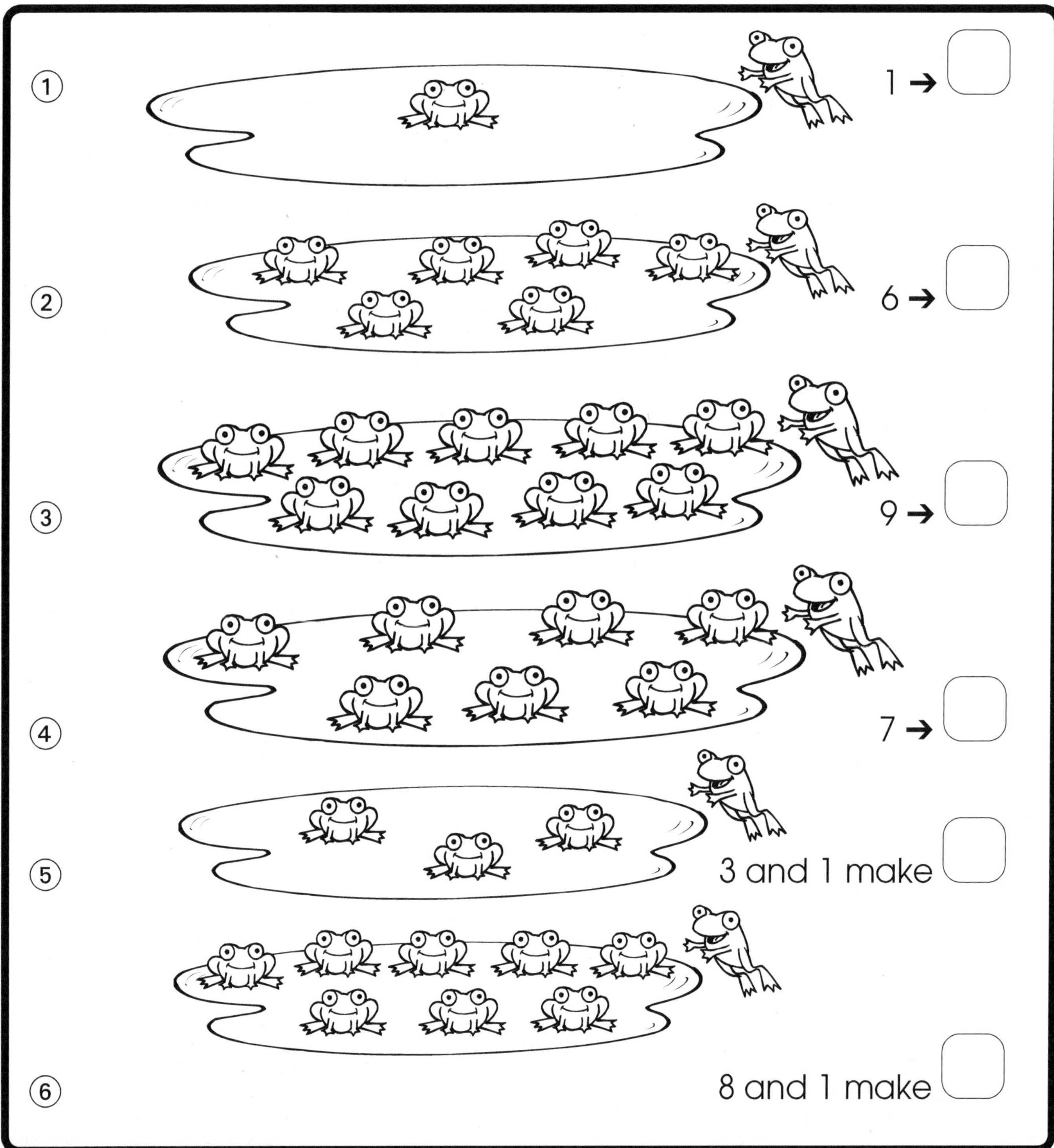

Count up 1

Adding is easy!
Adding is fun!
Circle the larger number
And count up 1!

① Circle the larger number in each number story and count up one.

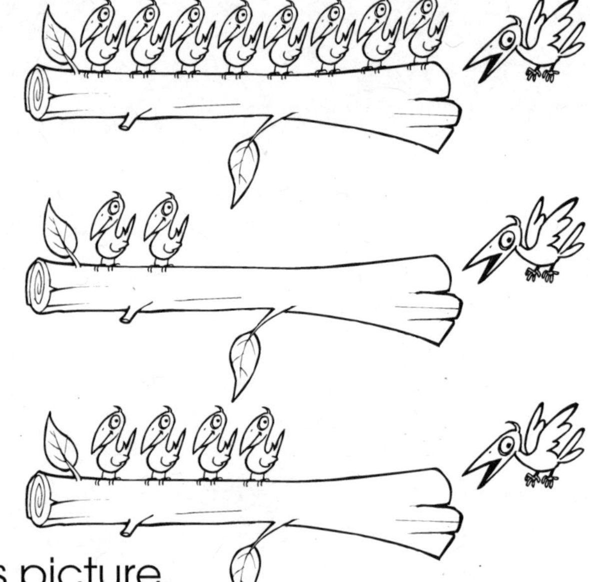

(a) 8 and 1 make ☐

(b) 2 and 1 make ☐

(c) 1 and 4 make ☐

② Match each number story to its picture.

(a) 1 and 6 make ☐ •   •

(b) 1 and 4 make ☐ •   •

(c) 5 and 1 make ☐ •   •

(d) 1 and 7 make ☐ •   •

Count up 2

When you know how
Adding's easy to do!
Add these numbers
By counting up 2!

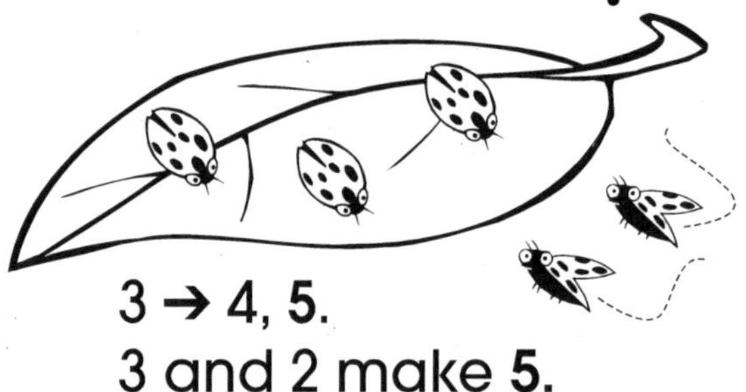

3 → 4, 5.
3 and 2 make **5.**

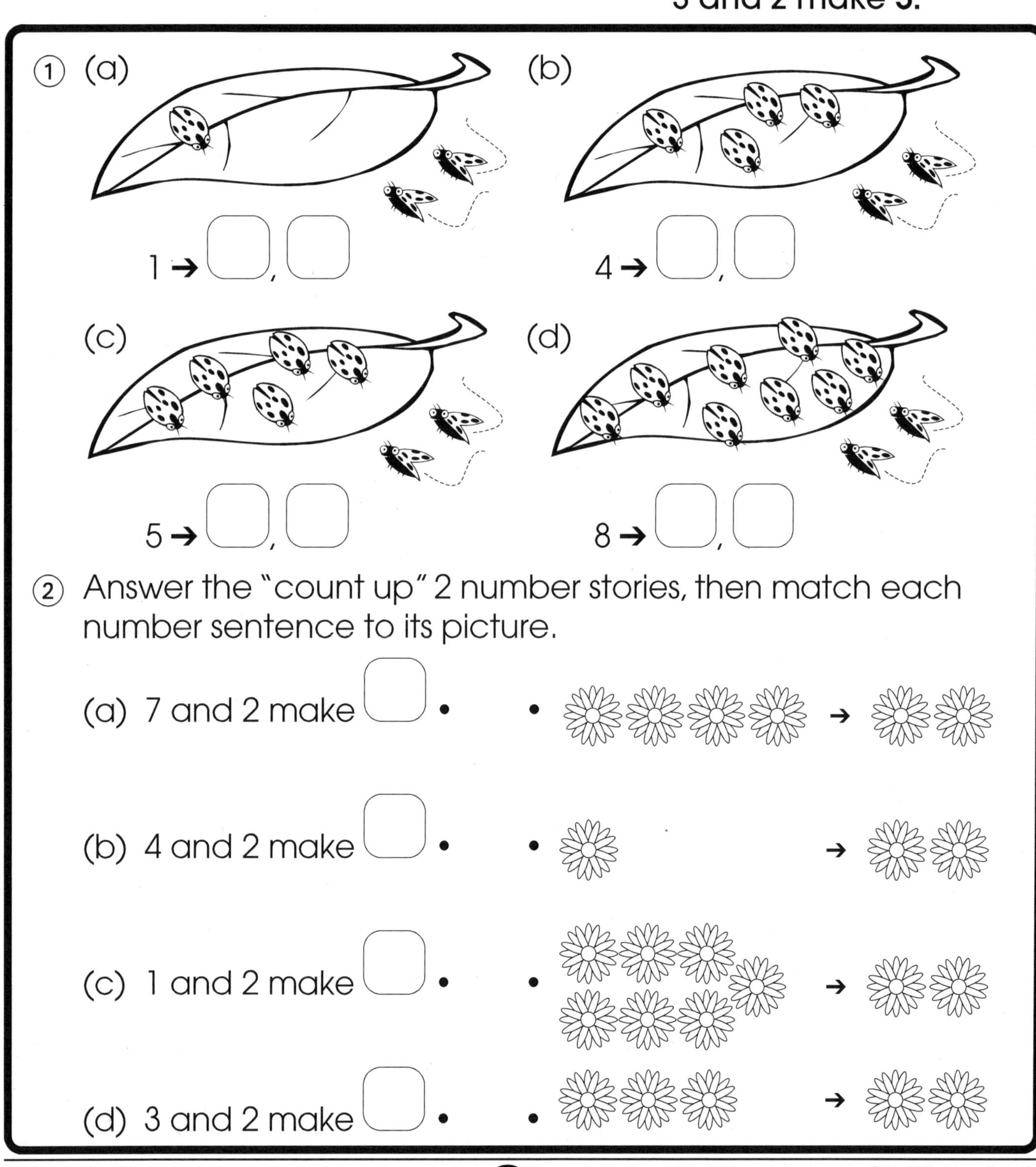

1. (a) 1 → ☐, ☐

   (b) 4 → ☐, ☐

   (c) 5 → ☐, ☐

   (d) 8 → ☐, ☐

2. Answer the "count up" 2 number stories, then match each number sentence to its picture.

   (a) 7 and 2 make ☐

   (b) 4 and 2 make ☐

   (c) 1 and 2 make ☐

   (d) 3 and 2 make ☐

When you know how
Adding's easy to do!
Circle the larger number
And count up 2!

Count up 2

4 → 5, **6**.
4 and 2 make **6**.

① Circle the larger number in each number story and count up 2.

(a) 2 and 3 make ☐ (b) 7 and 2 make ☐

(c) 8 and 2 make ☐ (d) 2 and 6 make ☐

(e) 2 and 5 make ☐ (f) 3 and 2 make ☐

② Circle the larger number and count up 1 or 2 to answer the number stories in the balloons.

(a) 2 and 6 make ☐

(b) 1 and 9 make ☐

(c) 7 and 1 make ☐

(d) 2 and 7 make ☐

(e) 5 and 2 make ☐

(f) 4 and 1 make ☐

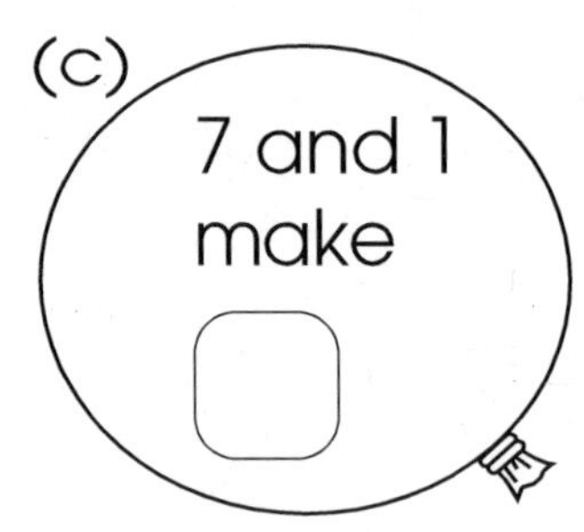

③ Color the "count up 1" number stories green and the "count up 2" number stories yellow.

Count up 3

4 ➔ 5, 6, **7**.
4 and 3 make **7**.

① Write the count up numbers and finish the number sentence.

(a) **6** ➔ ☐, ☐, ☐.

6 and **3** make ☐.

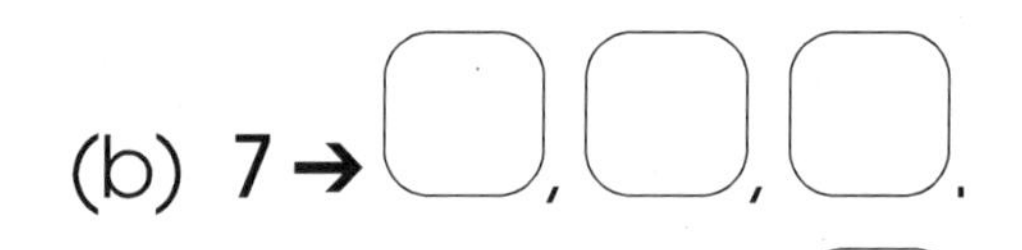

(b) **7** ➔ ☐, ☐, ☐.

7 and **3** make ☐.

(c) **5** ➔ ☐, ☐, ☐.

5 and **3** make ☐.

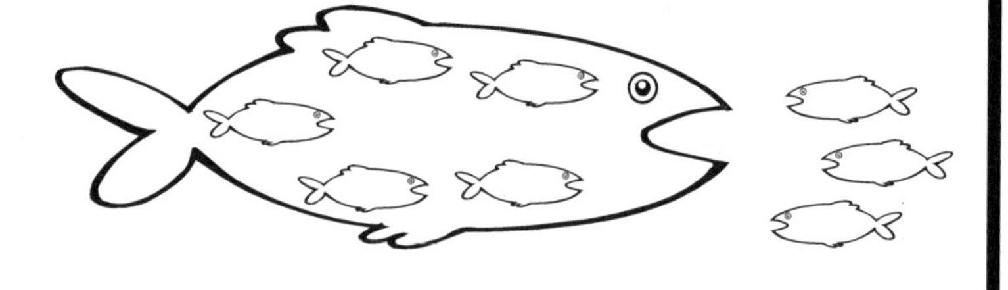

② Complete these "count up 3" number stories then match each number story to its picture.

(a) **2** and **3** make ☐ • •

(b) **6** and **3** make ☐ • •

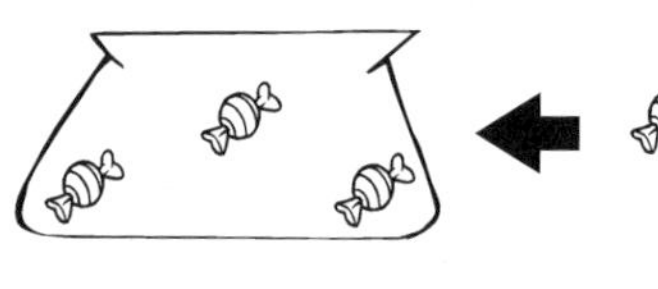

(c) **3** and **3** make ☐ • •

Count up 3

Adding is simple
For you and for me!
Circle the larger number
And count up 3!

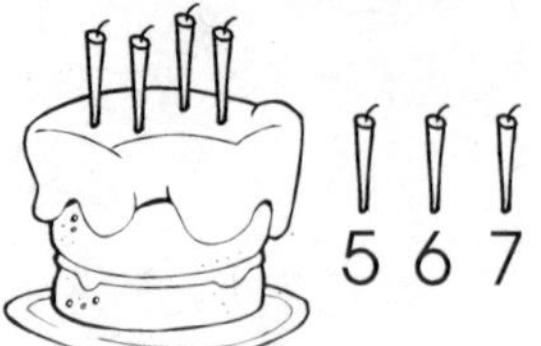

4 → 5, 6, **7**.
4 and 3 make **7**.

① Circle the larger number in each problem and count up 3.

(a) 5 and 3 make ☐

(b) 3 and 4 make ☐

(c) 6 and 3 make ☐

(d) 7 and 3 make ☐

② Circle the larger number and count up 1, 2, or 3 to answer the number stories in the rockets.

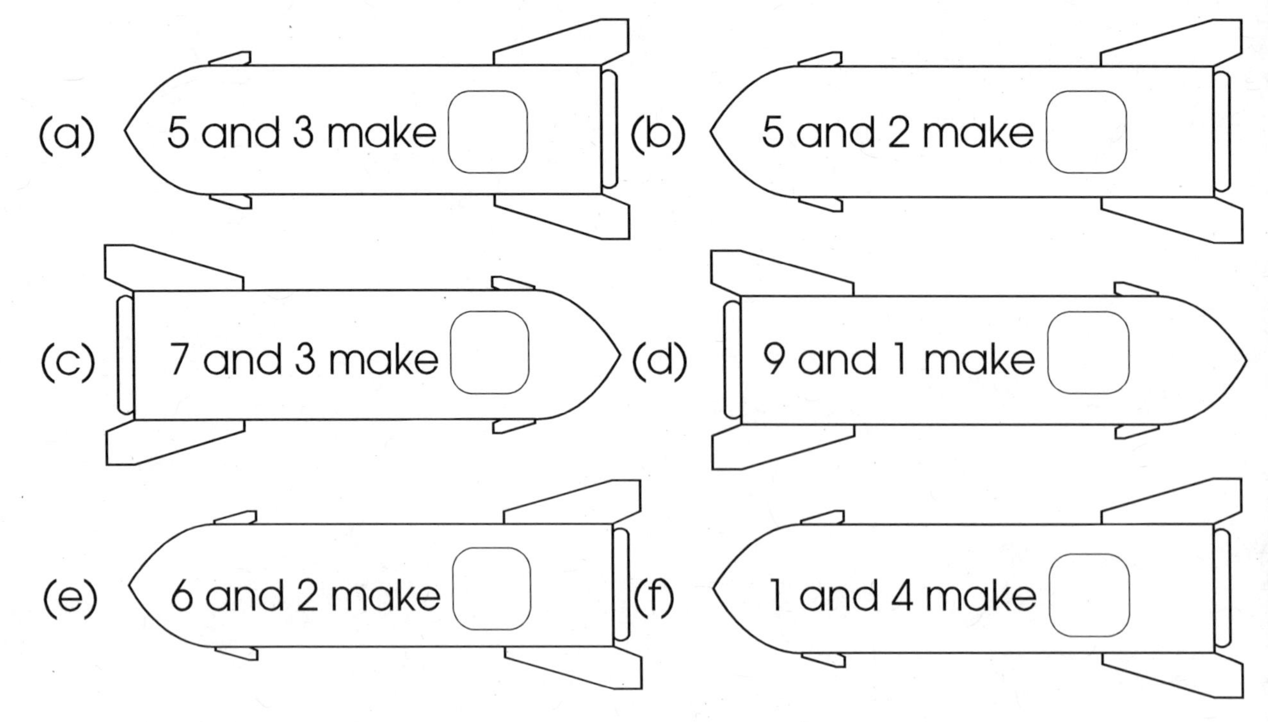

③ Color the "count up 1" number stories red, the "count up 2" number stories purple and the "count up 3" number stories blue.

Equal

"Equal" is a special word
for when things are the same.
It also has a special sign,
So don't forget its name!

① Color the "equal" sign.

② Color the beetles with an equal number of spots on each wing.

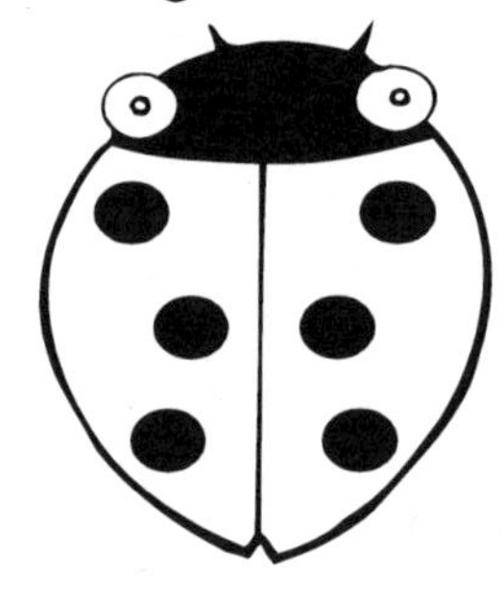

③ Draw spots on the wings of these beetles to make them equal.

④ Write an equal sign in the box next to the group with the same number of beetles as there are on the leaf.

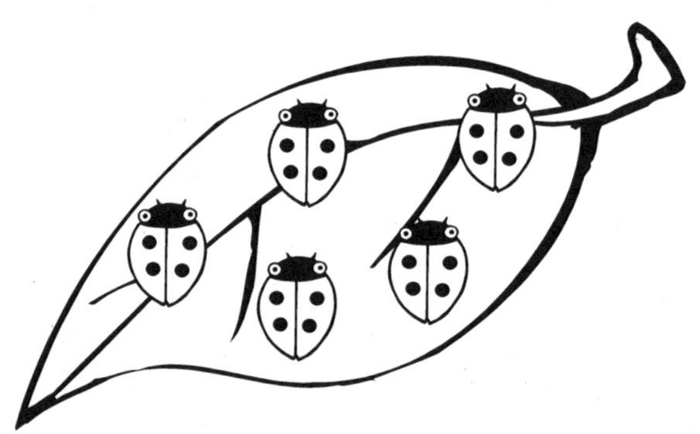

Adding 0

Adding zero is easy!
It's like a little game.
When you add zero,
The answer stays the same!

Do you know this sign?

It is a plus sign.
We can use it when we add numbers together.

① Write and draw a "zero" number story for each of the pictures.

+

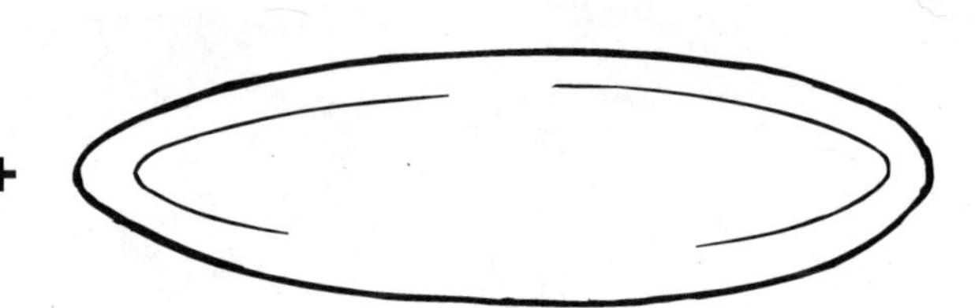

(a) ☐ muffins + ☐ muffins make ☐ muffins altogether.

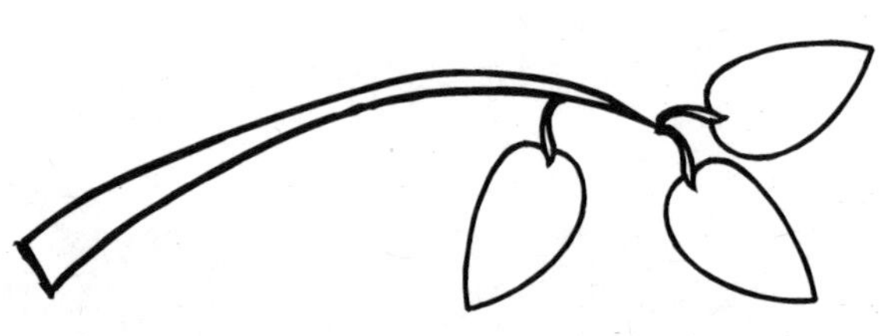

+

(b) ☐ leaves + ☐ leaves make ☐ leaves altogether.

② Write the missing numbers in these "zero" number stories.

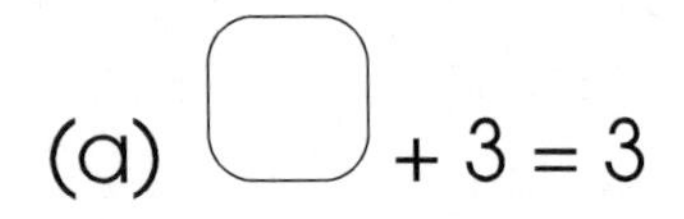

(a) ☐ + 3 = 3 (b) 6 + ☐ = 6

(c) 0 + ☐ = 4 (d) ☐ + 9 = 9

(e) ☐ + 4 = 4 (f) 5 + ☐ = 5

③ Adding zero is easy! How fast can you add these?

(a) 1 + 0 = ☐ (b) 0 + 4 = ☐ (c) 0 + 7 + ☐

Name

"Count up" Race

Reviewing addition to 10

Find the answers to these number stories. Write your time in the "finish" starburst and color your score on the scale.

Start

① 5 + 5 = ☐ ② 6 + 3 = ☐

④ 9 + 1 = ☐ ③ 8 + 2 = ☐

⑤ 6 + 4 = ☐ ⑥ 3 + 5 = ☐

⑧ 1 + 8 = ☐ ⑦ 2 + 7 = ☐

⑨ 7 + 3 = ☐ ⑩ 6 + 2 = ☐

⑫ 4 + 6 = ☐ ⑪ 7 + 1 = ☐

⑬ 3 + 4 = ☐ ⑭ 2 + 5 = ☐

⑯ 5 + 5 = ☐ ⑮ 1 + 1 = ☐

⑰ 4 + 4 = ☐ ⑱ 3 + 3 = ☐

⑳ 6 + 1 = ☐ ⑲ 5 + 2 = ☐

Finish

20
19
18
17
16
15
14
13
12
11
10
9
8
7
6
5
4
3
2
1
0

You're Joking!
Reviewing numbers to 20

Name

Answer these "count up" number sentences to solve the riddle.

What do you call a vegetarian vampire?

A 6 + 0 = ☐ **B** 9 + 3 = ☐ **C** 2 + 2 = ☐

D 15 + 3 = ☐ **E** 15 + 2 = ☐ **F** 0 + 5 = ☐

G 2 + 18 = ☐ **H** 12 + 7 = ☐ **I** 3 + 8 = ☐

J 16 + 3 = ☐ **K** 9 + 1 = ☐ **L** 3 + 15 = ☐

M 2 + 16 = ☐ **N** 2 + 11 = ☐ **O** 1 + 13 = ☐

P 2 + 12 = ☐ **Q** 0 + 8 = ☐ **R** 3 + 4 = ☐

S 3 + 10 = ☐ **T** 3 + 0 = ☐ **U** 9 + 1 = ☐

V 12 + 3 = ☐ **W** 3 + 13 = ☐ **X** 7 + 2 = ☐

Y 17 + 2 = ☐ **Z** 14 + 3 = ☐

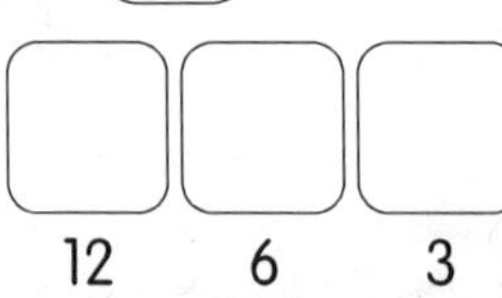

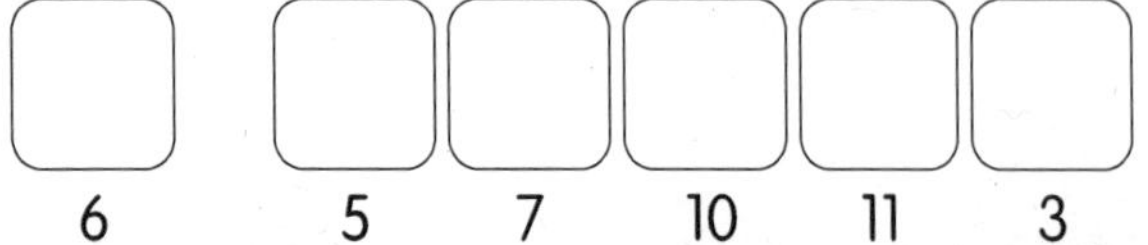

6   5 7 10 11 3   12 6 3

Swap the numbers
to and fro,
To double up
The facts you know!

Swap-around Facts

Count up 1

For example 1 + 6 = 7
is the same as 6 + 1 = 7

① Join the "swap-around" partners to answer these "count up 1" facts.

1 + 9 = ☐ • • 6 + 1 = ☐

1 + 6 = ☐ • • 9 + 1 = ☐

1 + 2 = ☐ • • 2 + 1 = ☐

② Color each "count up 1" number sentence the same color as its "swap-around" partner and then answer them.

1 + 4 = ☐		8 + 1 = ☐
6 + 1 = ☐	1 + 8 = ☐	1+ 5 = ☐
5 + 1 = ☐	1 + 6 = ☐	4 + 1 = ☐

③ Write the "swap-around" facts for these "count up 1" addition problems.

3 + 1 = ☐ → ☐ + ☐ = ☐

1 + 8 = ☐ → ☐ + ☐ = ☐

Swap-around Facts

Count up 2

Swap the numbers
to and fro,
To double up
The facts you know!

For example, **2 + 5 = 7**
is the same as **5 + 2 = 7**

① Join the "swap-around" partners to answer these "count up 2" facts.

2 + 8 = ☐ •	• 6 + 2 = ☐
2 + 6 = ☐ •	• 3 + 2 = ☐
2 + 3 = ☐ •	• 8 + 2 = ☐

② Color each "count up 2" number sentence the same color as its "swap-around" partner and then answer them.

2 + 4 = ☐		8 + 2 = ☐
2 + 5 = ☐	7+ 2 = ☐	2 + 7 = ☐
4 + 2 = ☐	2 + 8 = ☐	5 + 2 = ☐

③ Write the "swap-around" facts for these "count up 2" addition problems.

4 + 2 = ☐ ➔ ☐ + ☐ = 6

7 + 2 = ☐ ➔ ☐ + ☐ = 9

Swap the numbers
to and fro,
To double up
The facts you know!

Swap-around Facts

Count up 3

For example, $3 + 6 = 9$
is the same as $6 + 3 = 9$

① Join the "swap-around" partners to answer these "count up 3" facts.

$3 + 4 = \square$ • • $7 + 3 = \square$

$3 + 6 = \square$ • • $6 + 3 = \square$

$3 + 7 = \square$ • • $4 + 3 = \square$

② Color each "count up 3" number sentence the same color as its "swap-around" partner and write the answer.

$3 + 5 = \square$		$7 + 3 = \square$
$6 + 3 = \square$	$4 + 3 = \square$	$5 + 3 = \square$
$3 + 7 = \square$	$3 + 6 = \square$	$3 + 4 = \square$

③ Write the "swap-around" facts for these "count up 3" addition problems.

$3 + 6 = \square$ ➔ $\square + \square = 9$

$3 + 8 = \square$ ➔ $\square + \square = 11$

"Swap-around" Street

① Answer the "swap-around" facts by circling the larger number and counting up 1, 2, or 3.

(a) 1 + 2 = ☐	(b) 2 + 3 = ☐	(c) 3 + 4 = ☐
(d) 1 + 5 = ☐	(e) 2 + 6 = ☐	(f) 3 + 7 = ☐
(g) 1 + 8 = ☐	(h) 2 + 4 = ☐	(i) 3 + 5 = ☐
(j) 1 + 9 = ☐	(k) 2 + 7 = ☐	(l) 3 + 6 = ☐

(m) 2 + 1 = ☐	(n) 4 + 2 = ☐	(o) 4 + 3 = ☐
(p) 5 + 1 = ☐	(q) 7 + 2 = ☐	(r) 5 + 3 = ☐
(s) 8 + 1 = ☐	(t) 3 + 2 = ☐	(u) 7 + 3 = ☐
(v) 9 + 1 = ☐	(w) 6 + 2 = ☐	(x) 6 + 3 = ☐

② Color each number sentence the same color as its swap-around partner.

③ Cut out the bricks and roofs and glue the swap-around partners together on a separate sheet of paper to make a street.

Doubles – 1, 2, 3

① Use the pictures to help you answer these "double" facts.

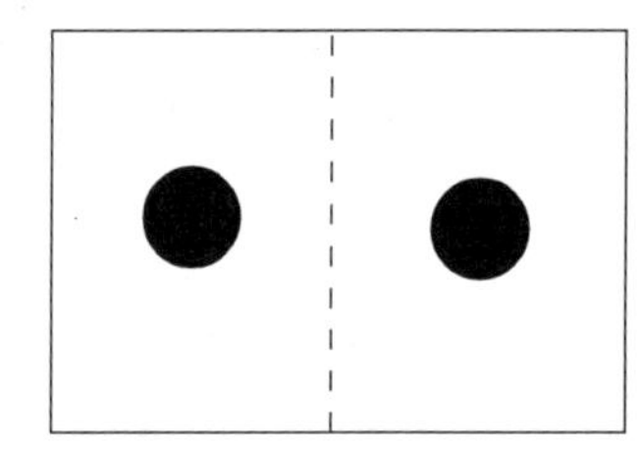

(a) 1 + 1 = ☐

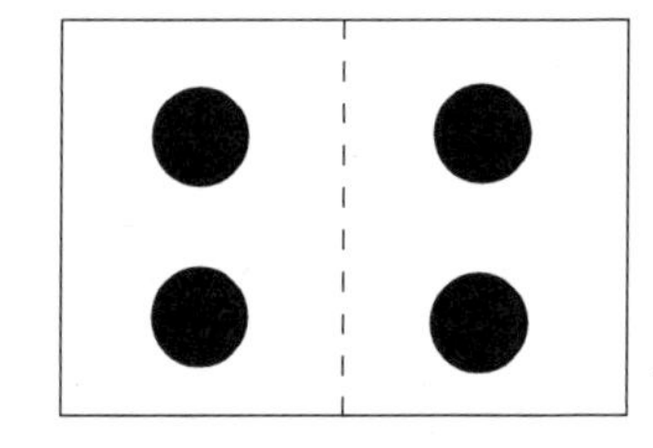

(b) 2 + 2 = ☐

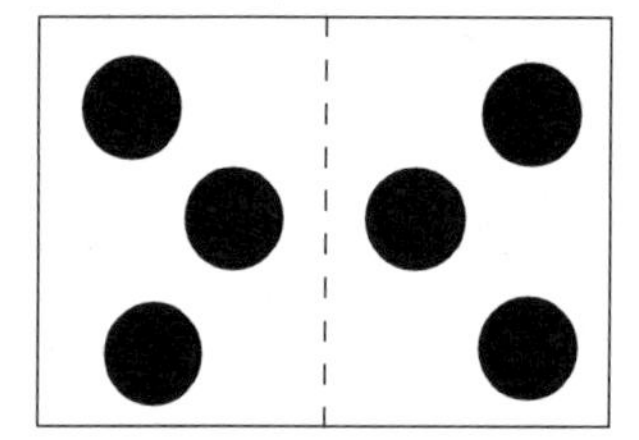

(c) 3 + 3 = ☐

② Finish these pictures to make them double. Complete the number stories.

(a)

2 ducks and 2 ducks make ☐ ducks.

(b)

3 daisies and 3 daisies make ☐ daisies.

(c)

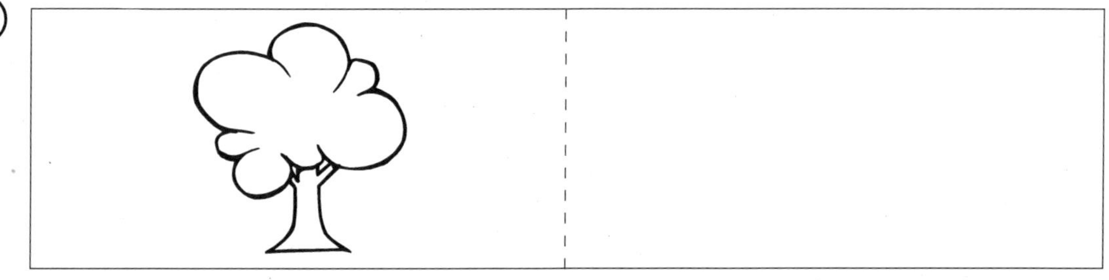

1 tree and 1 tree make ☐ trees.

Double the pictures to fill the fish tank.

(a) Double 1 octopus → ☐ octopuses

(b) Double 2 fish → ☐ fish

(c) Double 3 shells → ☐ shells

Doubles – 4, 5, 6

① Use the pictures to help you answer these "double" facts.

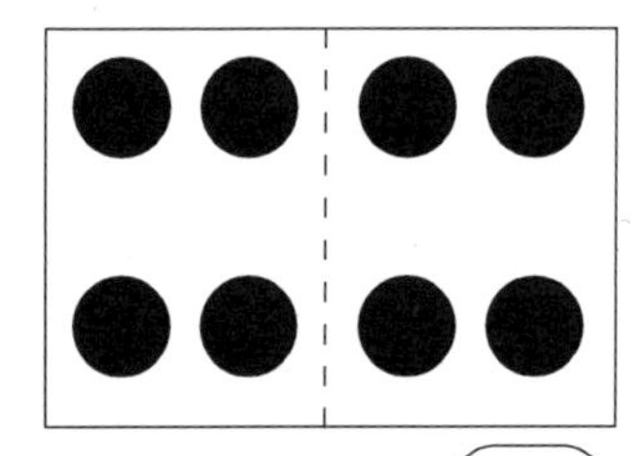

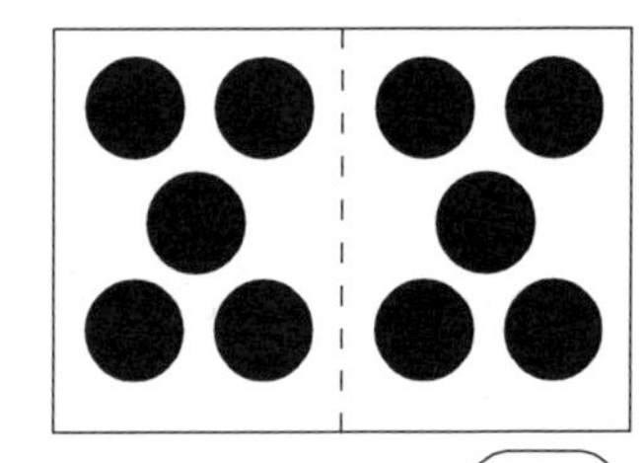

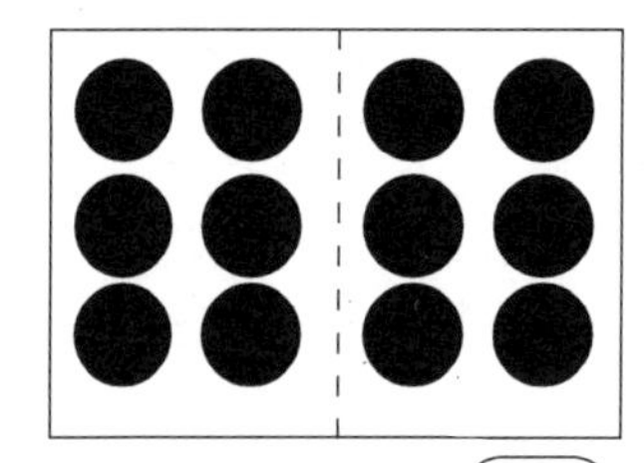

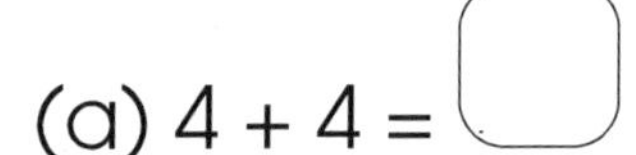

(a) 4 + 4 = ☐ (b) 5 + 5 = ☐ (c) 6 + 6 = ☐

② Finish these pictures to make them double. Complete the number stories.

(a)

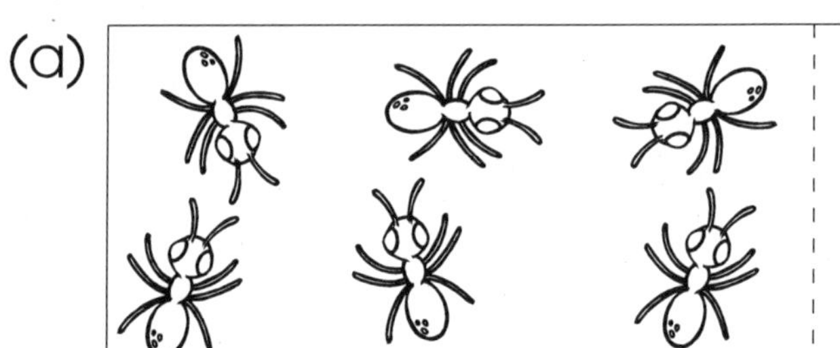

6 ants and 6 ants make ☐ ants.

(b)

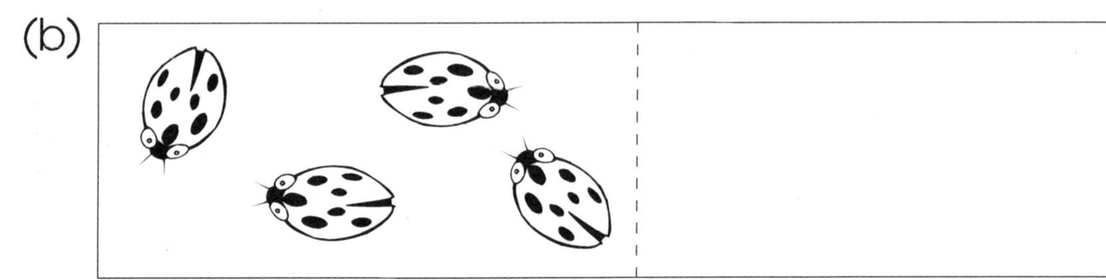

4 beetles and 4 beetles make ☐ beetles.

(c)

5 snails and 5 snails make ☐ snails.

Doubles – 4, 5, 6

Double the pictures to fill the lunch box.

(a) Double 4 rolls ➔ ☐ rolls

(b) Double 5 apples ➔ ☐ apples

(c) Double 6 drinks ➔ ☐ drinks

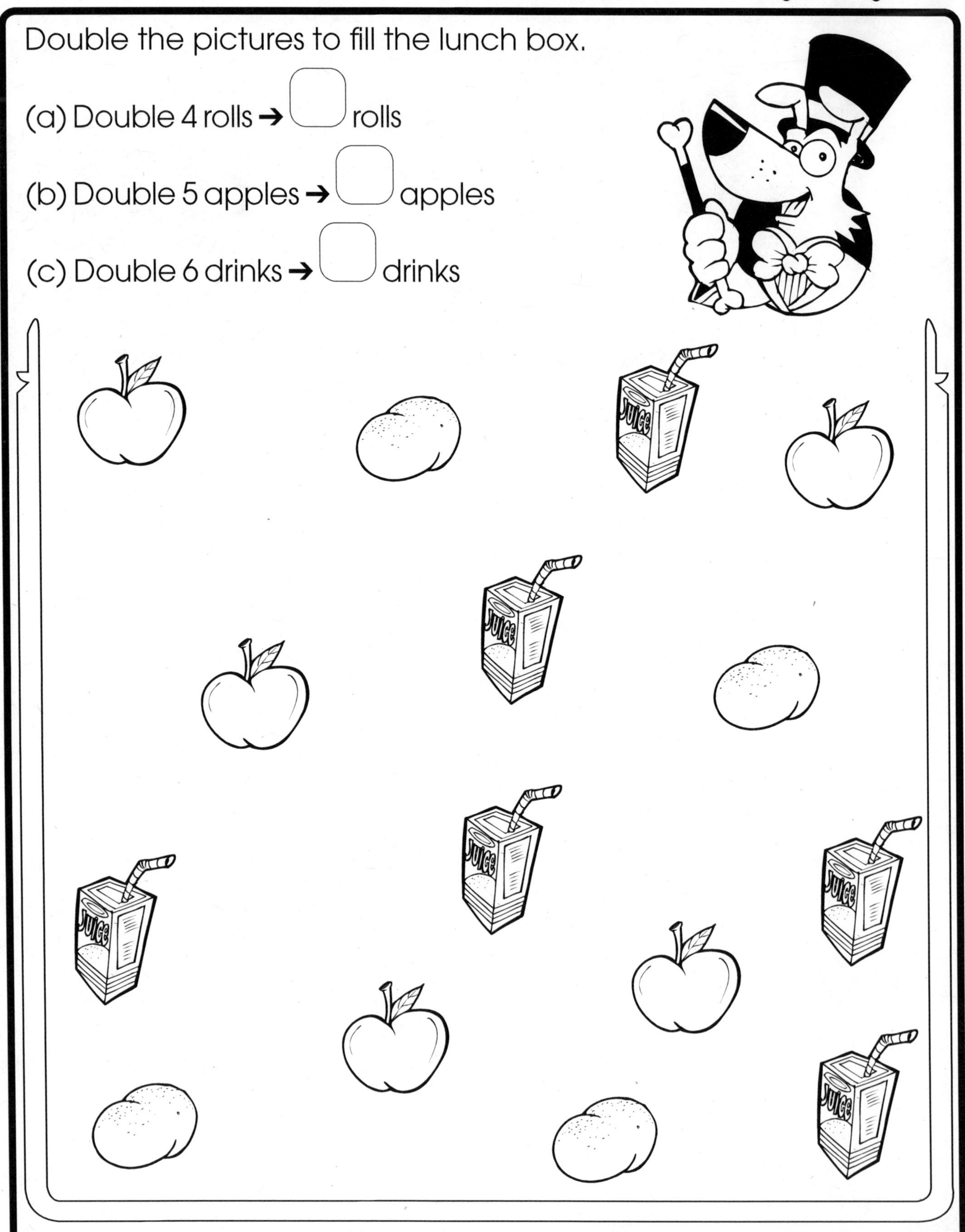

Doubles – 7, 8, 9

① Use the pictures to help you answer these "double" facts.

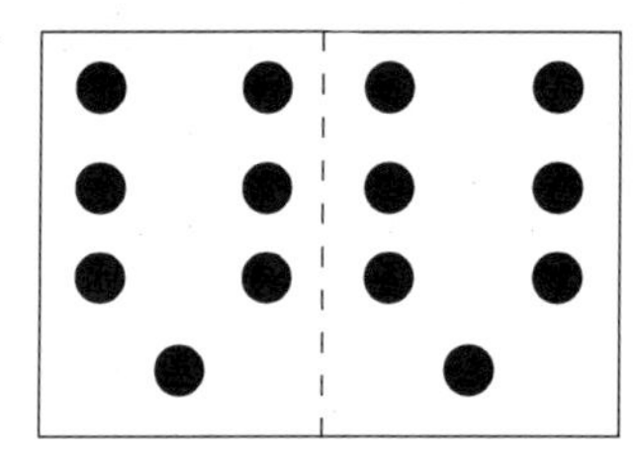

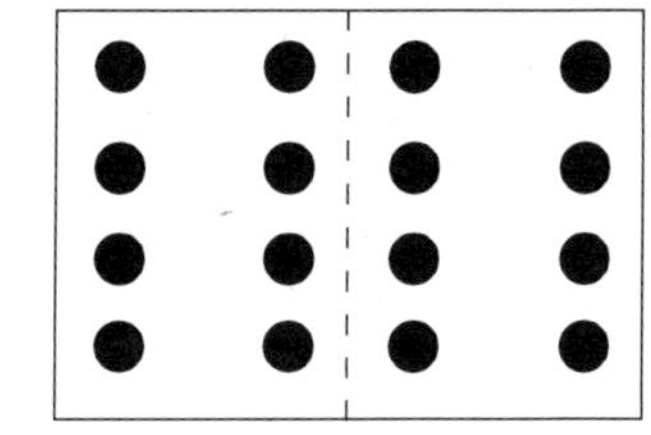

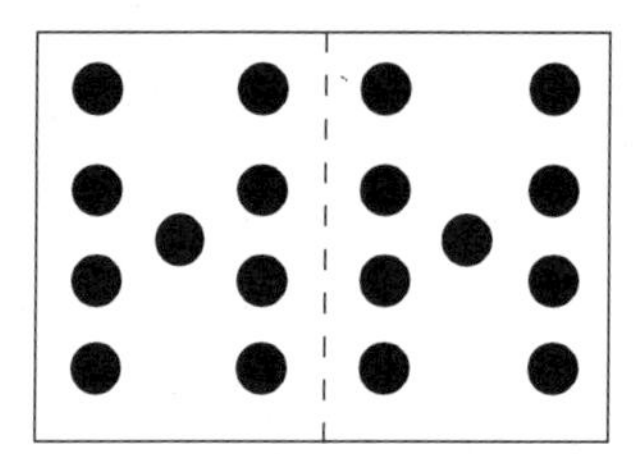

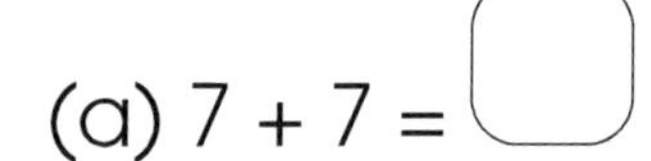

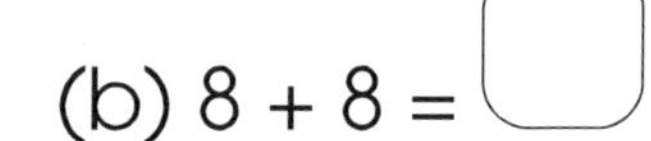

(a) 7 + 7 = ☐   (b) 8 + 8 = ☐   (c) 9 + 9 = ☐

② Finish these pictures to make them double. Complete the number stories.

(a)

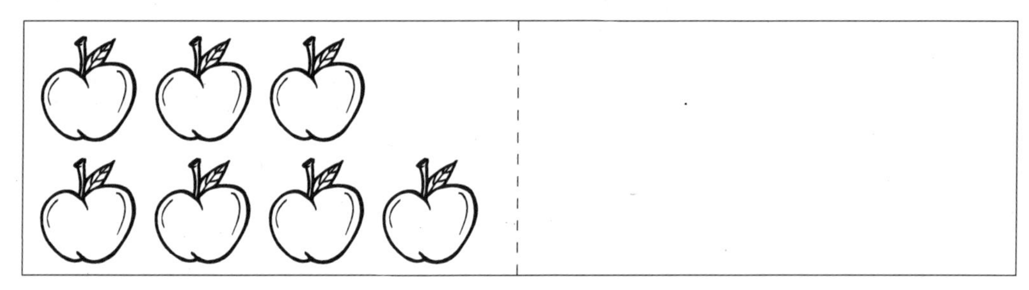

7 apples and 7 apples make ☐ apples.

(b)

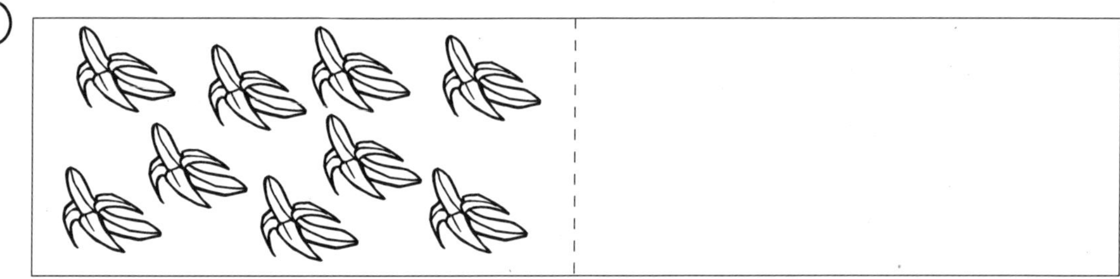

9 bananas and 9 bananas make ☐ bananas

(c)

8 grapes and 8 grapes make ☐ grapes.

Doubles – 7, 8, 9

Double the pictures to fill the picnic blanket.

(a) Double 7 hot dogs → ☐ hot dogs

(b) Double 8 cookies → ☐ cookies

(c) Double 9 candies → ☐ candies

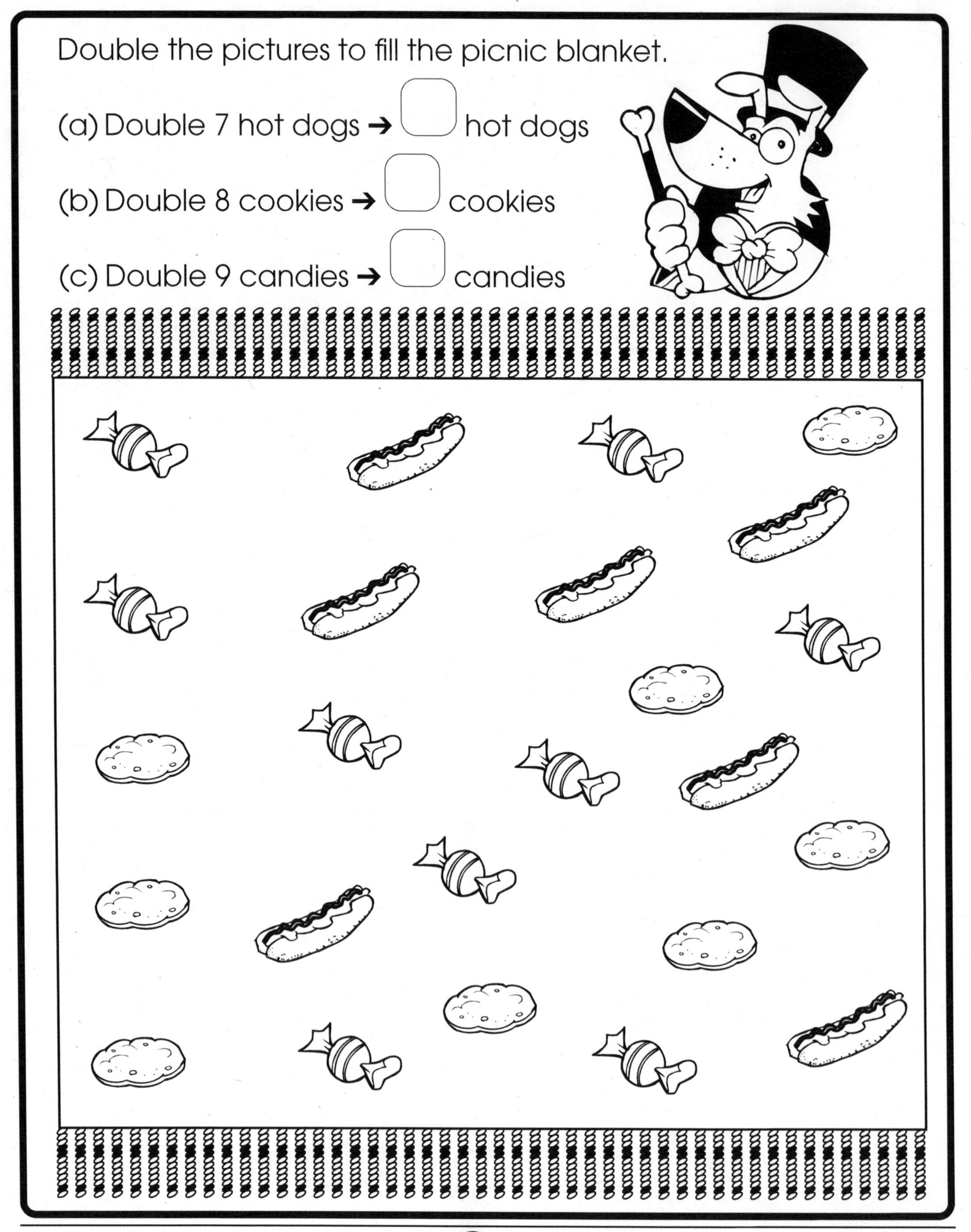

Double Trouble!

Match these pictures to a "double" fact. Use the pictures to help you answer the number sentences.

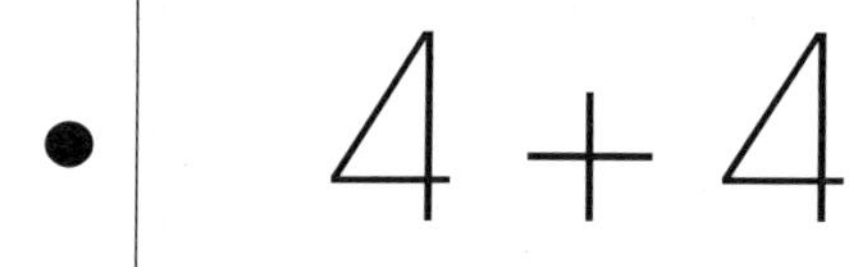	4 + 4
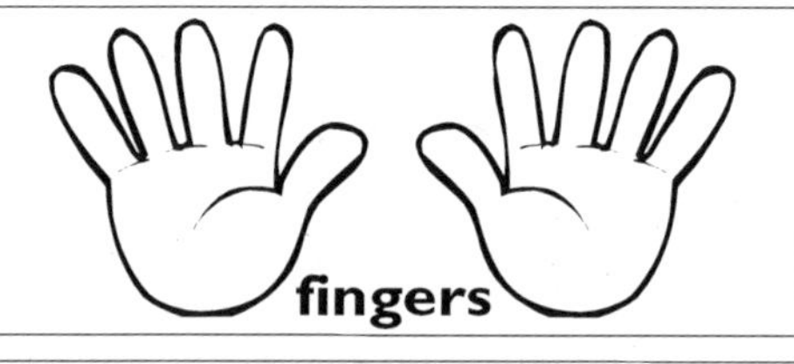	7 + 7
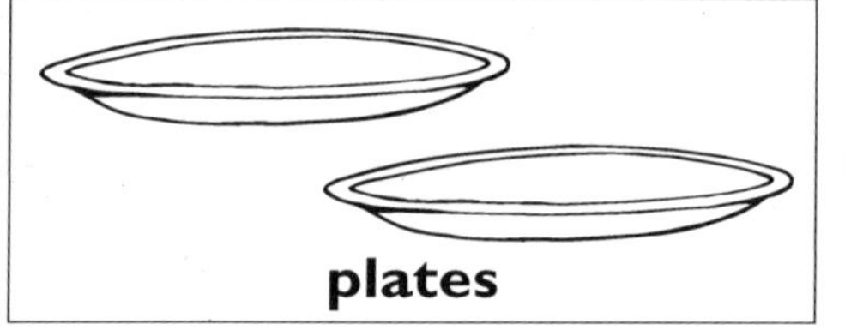	3 + 3
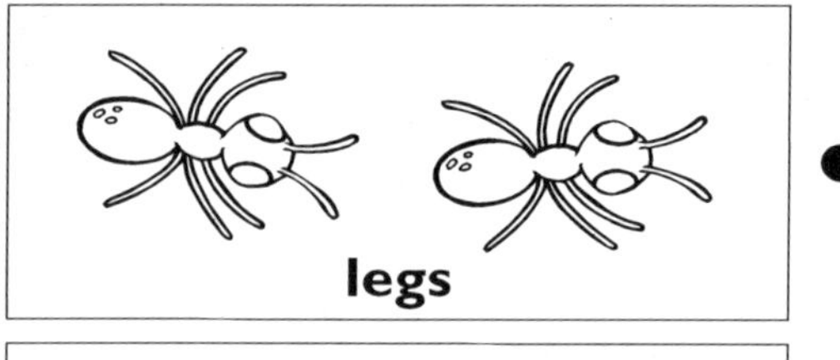	1 + 1
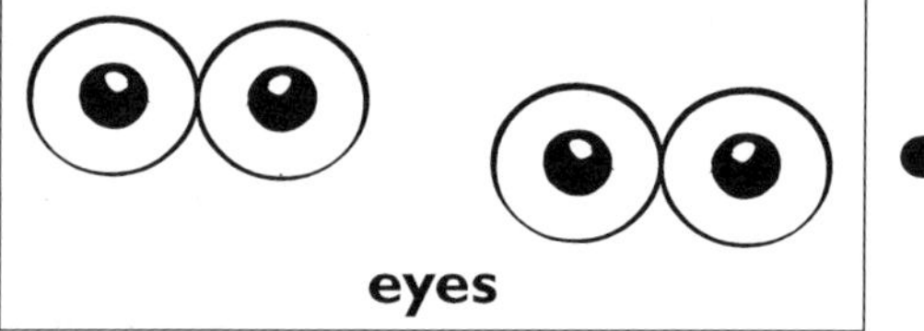	2 + 2
	5 + 5
	6 + 6
	8 + 8

Doubles – Review

① Add the doubles.

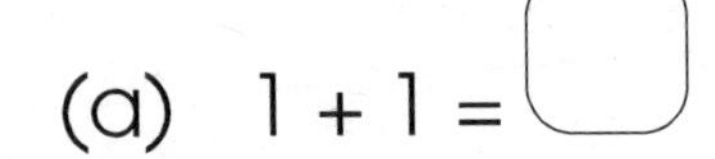

(a) 1 + 1 = ☐　(b) 2 + 2 = ☐　(c) 3 + 3 = ☐

(d) 4 + 4 = ☐　(e) 5 + 5 = ☐　(f) 6 + 6 = ☐

(g) 7 + 7 = ☐　(h) 8 + 8 = ☐　(i) 9 + 9 = ☐

② Count the number of diamonds on each card. Double the number of diamonds in your head to help you complete the sum. For example:

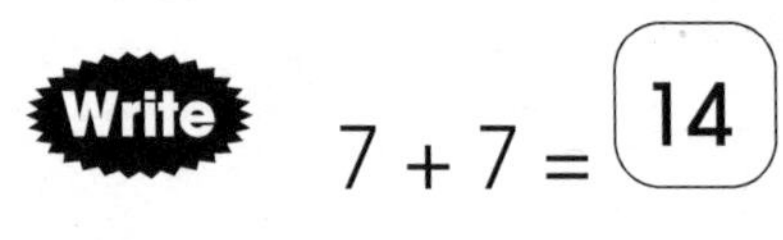

(a) Double ☐

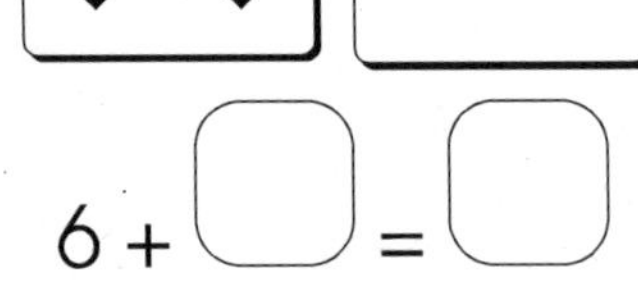

6 + ☐ = ☐

(b) Double ☐

9 + ☐ = ☐

(c) Double ☐

3 + ☐ = ☐

(d) Double ☐

5 + ☐ = ☐

(e) Double ☐

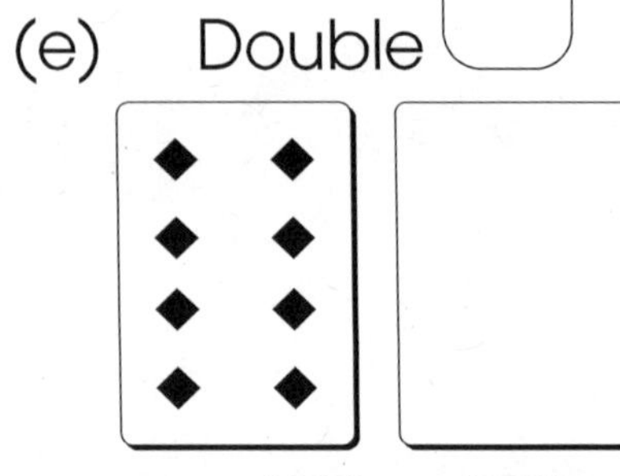

8 + ☐ = ☐

(f) Double ☐

4 + ☐ = ☐

Doubles – Review

Answer the number sentences. Color the "doubles" facts triangles yellow.

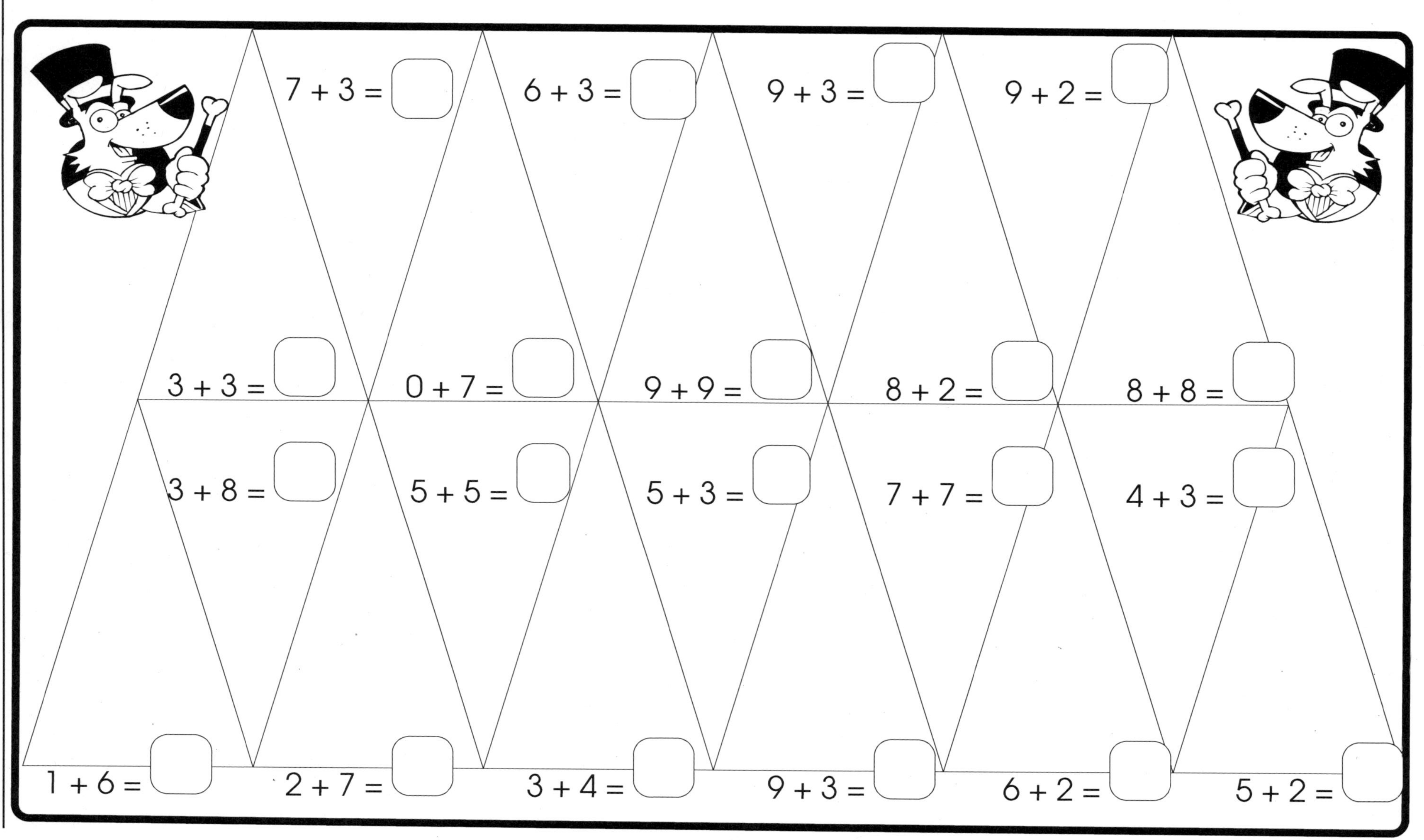

Double and Count up 1

Double the small one
And then count up
To solve the problem
You're working on!

2 + 3 = 5

Think 2 + 2 + 1

Double 2 and count up 1

Color the squares to show the double and complete the "think" boxes to help you answer each problem.

Remember ... double the smaller number and count up!

① 5 + 6 = ☐ **Think** Double ☐ and count up ☐

② 8 + 7 = ☐ **Think** Double ☐ and count up ☐

③ 4 + 5 = ☐ **Think** Double ☐ and count up ☐

④ 7 + 6 = ☐ **Think** Double ☐ and count up ☐

⑤ 9 + 8 = ☐ **Think** Double ☐ and count up ☐

Doubles and Doubles Count up 1

Answer the number sentences.
Color the "doubles" boxes red.
Color the "doubles count up 1" boxes green.

1. 1 + 2 = ☐	11. 1 + 2 = ☐	21. 7 + 8 = ☐
2. 6 + 5 = ☐	12. 6 + 7 = ☐	22. 5 + 6 = ☐
3. 4 + 5 = ☐	13. 7 + 6 = ☐	23. 5 + 4 = ☐
4. 9 + 8 = ☐	14. 4 + 4 = ☐	24. 7 + 6 = ☐
5. 3 + 3 = ☐	15. 9 + 9 = ☐	25. 6 + 6 = ☐
6. 8 + 9 = ☐	16. 5 + 5 = ☐	26. 1 + 2 = ☐
7. 9 + 8 = ☐	17. 5 + 6 = ☐	27. 4 + 3 = ☐
8. 6 + 7 = ☐	18. 3 + 2 = ☐	28. 8 + 7 = ☐
9. 4 + 5 = ☐	19. 6 + 5 = ☐	29. 8 + 9 = ☐
10. 2 + 3 = ☐	20. 3 + 4 = ☐	30. 2 + 1 = ☐

What is the sign colored red? ____________________

"Magic 10" Facts

"Magic 10" facts add to 10,
That's the only way they end!
If you're clever, soon you'll see,
You'll be a "math-magician" like me!

① Learn the facts! Then swap them around!

1 + 9 = ☐ 2 + 8 = ☐ 3 + 7 = ☐ 4 + 6 = ☐

9 + 1 = ☐ 8 + 2 = ☐ 7 + 3 = ☐ 6 + 4 = ☐

Double 5 is a special case! 5 + 5 = ☐

② Use the number lines to help you "count up" to find the missing number in these "magic 10" number sentences.

For example, 6 + ☐ = 10

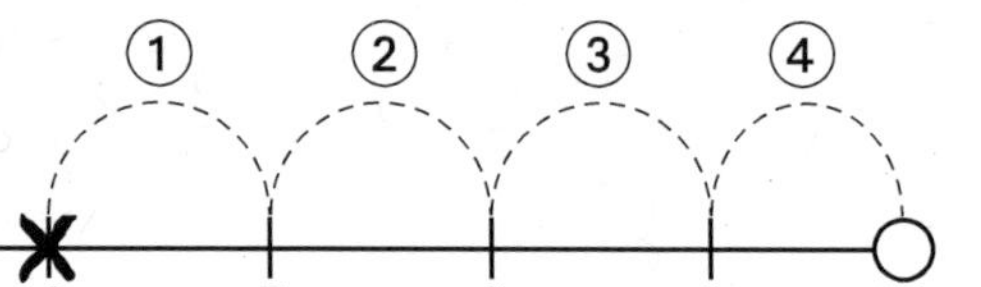

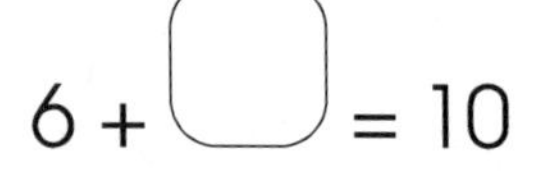

(a) 8 + ☐ = 10

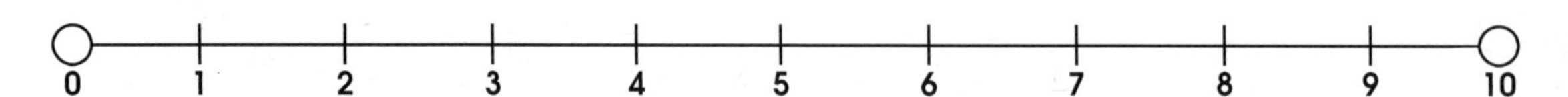

(b) 3 + ☐ = 10

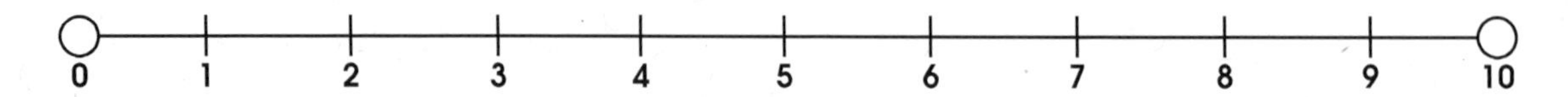

(c) 5 + ☐ = 10

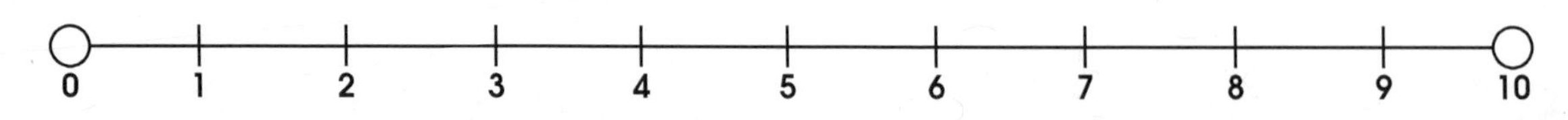

"Magic 10" Facts

① Complete the "magic 10" number facts.

(a) 1 + ☐ = 10 (b) 3 + ☐ = 10

(c) 4 + ☐ = 10 (d) 6 + ☐ = 10

(e) 7 + ☐ = 10 (f) 9 + ☐ = 10

(g) 1 + ☐ = 10 (h) 3 + ☐ = 10

(i) 4 + ☐ = 10 (j) 6 + ☐ = 10

② Color and cut out the numbers. Glue each pair of "magic 10" numbers back-to-back and attach them to the number 10 to make a mobile.

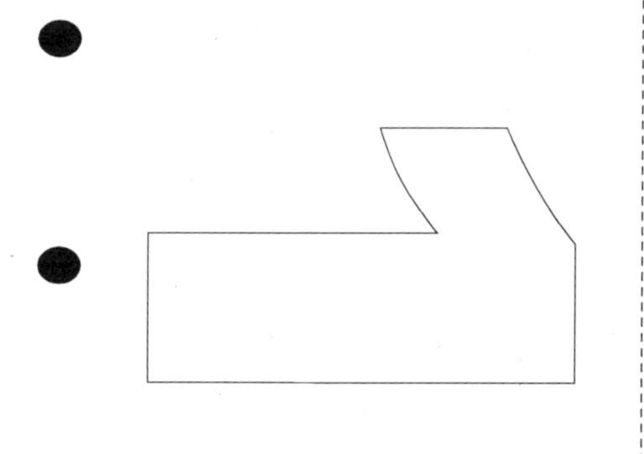

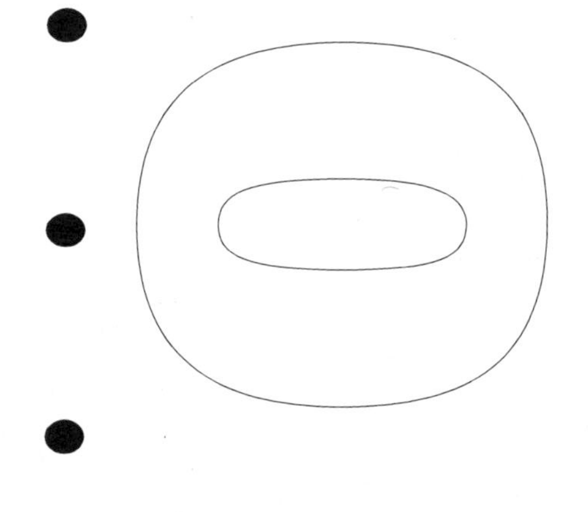

Using "Magic 10" Facts

Using "magic 10" facts
Is a clever game to play.
Add to 10 to find the answers
In another way!

① Circle groups of 10 and add on the "left-overs" to find the answers to these additions.

(a) 9 and 4 = ?

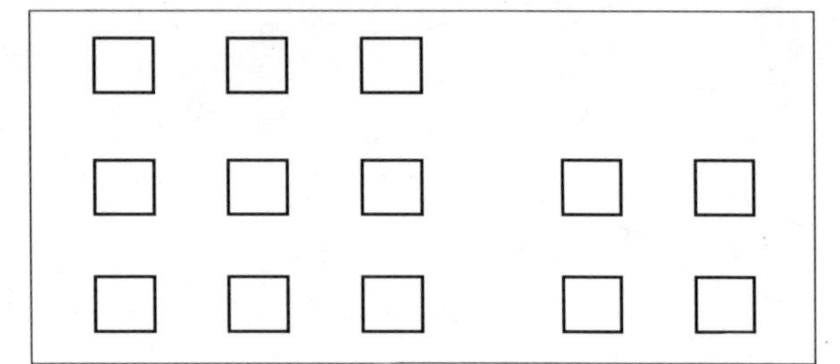

10 + ☐ = ☐

(b) 8 and 5 = ?

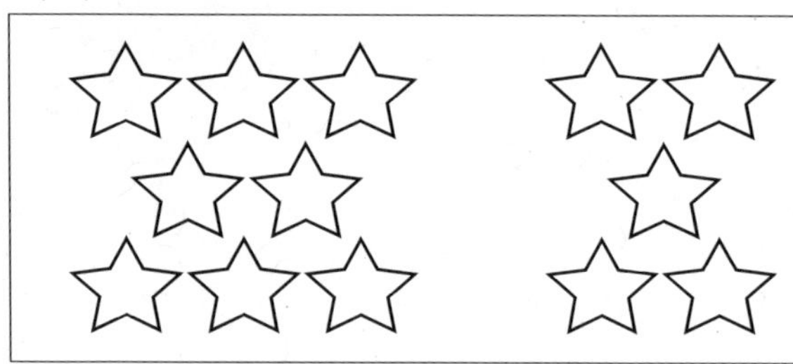

10 + ☐ = ☐

Try this! 

9 and 1 make 10
and 2 more make 12.

 See 9 + 3 = 12  Think 10 + 2 = 12

② Draw extra squares to help you complete the sums.

(a) 8 + 5 = ☐ 10 + ☐ = ☐ Think

(b) 6 + 5 = ☐ 10 + ☐ = ☐ Think

(c) 7 + 4 = ☐ 10 + ☐ = ☐ Think

(d) 9 + 6 = ☐ 10 + ☐ = ☐ Think

Use "magic 10" facts
to help your quest.
Count up to make 10
Then add the rest.

Building to 10

① Add 1 to 9 to make 10, then complete the number sentences.

(a) $9^{+1} + 4^{-1} =$ ☐ *becomes* 10 + ☐ = ☐

(b) $9^{+1} + 6^{-1} =$ ☐ *becomes* 10 + ☐ = ☐

(c) $9^{+1} + 3^{-1} =$ ☐ *becomes* 10 + ☐ = ☐

② Add 2 to 8 to make 10, then complete the number sentences.

(a) $8^{+2} + 5^{-2} =$ ☐ *becomes* 10 + ☐ = ☐

(b) $8^{+2} + 3^{-2} =$ ☐ *becomes* 10 + ☐ = ☐

(c) $8^{+2} + 6^{-2} =$ ☐ *becomes* 10 + ☐ = ☐

③ Add 3 to 7 to make 10, then complete the number sentences.

(a) $7^{+3} + 5^{-3} =$ ☐ *becomes* 10 + ☐ = ☐

(b) $7^{+3} + 8^{-3} =$ ☐ *becomes* 10 + ☐ = ☐

(c) $7^{+3} + 4^{-3} =$ ☐ *becomes* 10 + ☐ = ☐

④ Build the larger number to 10 in your head to help solve these number sentences.

(a) 8 + 4 = ☐ (b) 7 + 6 = ☐ (c) 9 + 2 = ☐

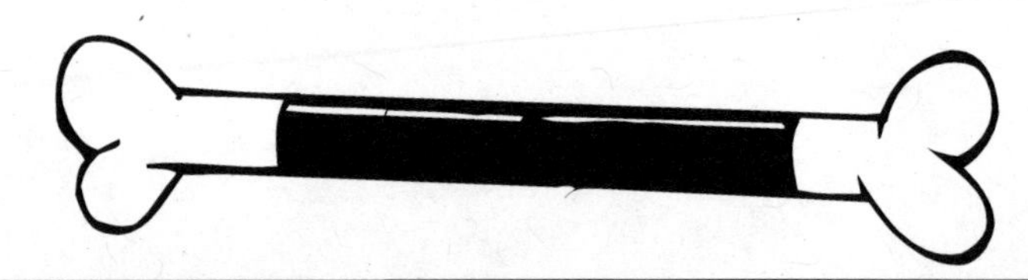

Adding on to 10

① Add on to 10.

(a) 10 + 3 = ☐

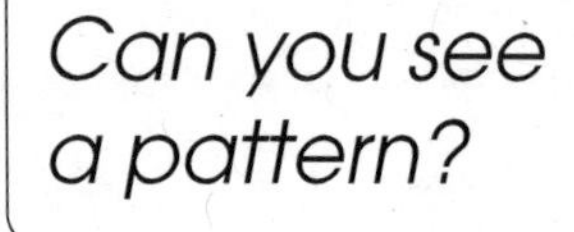

(b) 10 + 6 = ☐

(c) 10 + 9 = ☐

(d) 10 + 2 = ☐

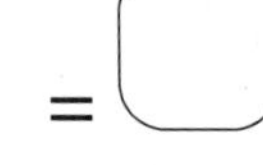

(e) 10 + 7 = ☐

② Use this pattern to complete these number stories.

(a) 10 + 1 = ☐ (b) 10 + 4 = ☐

(c) 10 + 8 = ☐ (d) 10 + 3 = ☐

③ Match these "swap around" facts to their partners and then complete the number sentences.

4 + 10 = ☐ •	• 10 + 1 = ☐
7 + 10 = ☐ •	• 10 + 6 = ☐
6 + 10 = ☐ •	• 10 + 4 = ☐
1 + 10 = ☐ •	• 10 + 9 = ☐
9 + 10 = ☐ •	• 10 + 7 = ☐

All the facts to 10

① Count up 1.

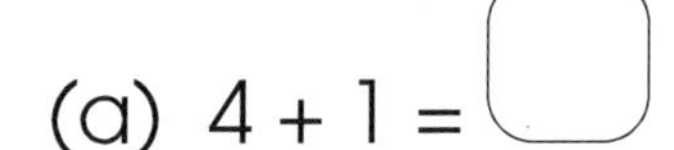

(a) 4 + 1 = ☐ (b) 9 + 1 = ☐ (c) 6 + 1 = ☐

(d) 1 + 8 = ☐ (e) 1 + 7 = ☐

How did you do?

1	2	3	4	5

② Count up 2.

(a) 3 + 2 = ☐ (b) 6 + 2 = ☐ (c) 2 + 4 = ☐

(d) 2 + 7 = ☐ (e) 5 + 2 = ☐

How did you do?

1	2	3	4	5

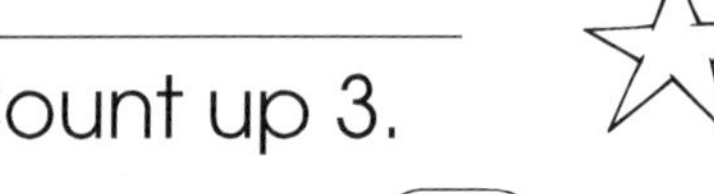

③ Count up 3.

(a) 5 + 3 = ☐ (b) 3 + 6 = ☐ (c) 4 + 3 = ☐

(d) 3 + 7 = ☐ (e) 3 + 5 = ☐

How did you do?

1	2	3	4	5

④ Magic 10 facts and doubles.

(a) 5 + 5 = ☐ (b) 6 + 4 = ☐ (c) 2 + 8 = ☐

(d) 2 + 2 = ☐ (e) 7 + 3 = ☐

How did you do?

1	2	3	4	5

⑤ Mixed up Magic.

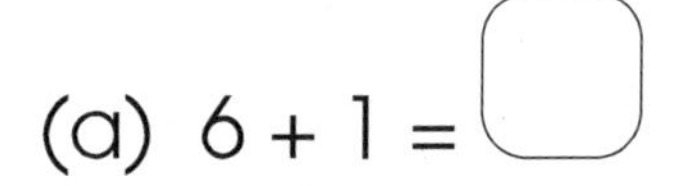

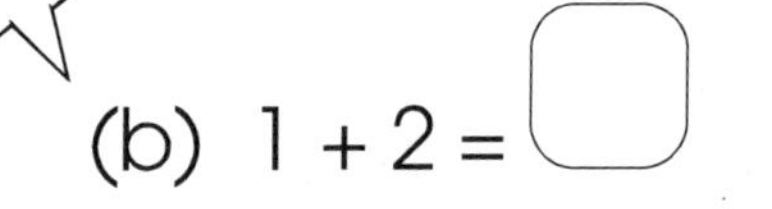

(a) 6 + 1 = ☐ (b) 1 + 2 = ☐ (c) 2 + 3 = ☐

(d) 2 + 2 = ☐ (e) 9 + 1 = ☐

How did you do?

1	2	3	4	5

All the facts to 10

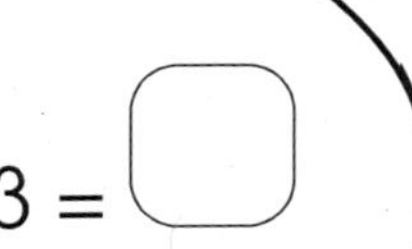
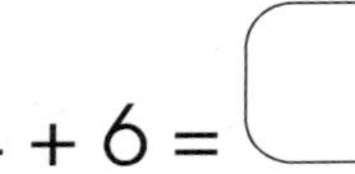
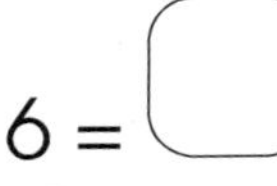
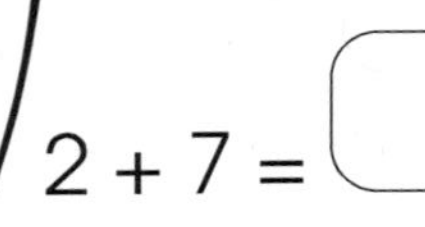

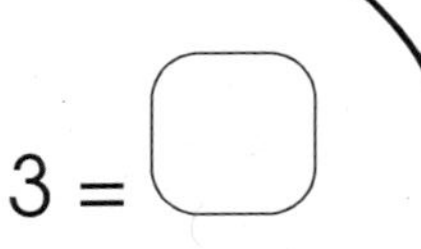
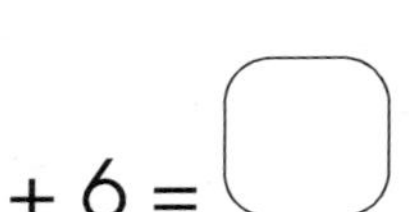

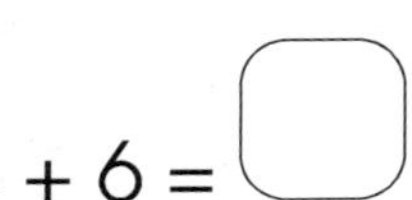
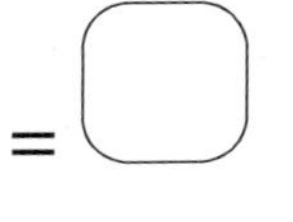
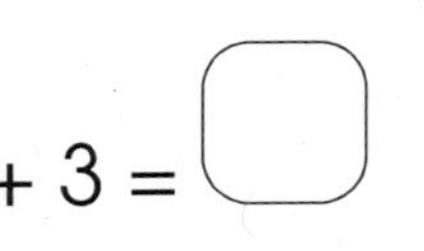

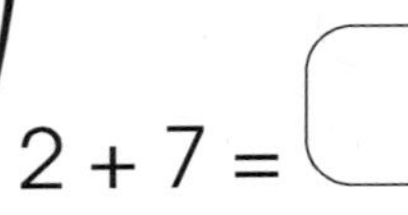
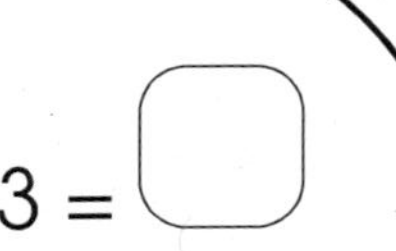
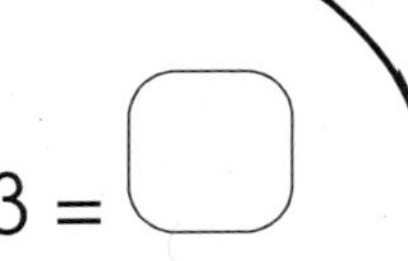
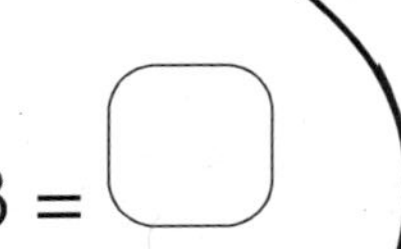
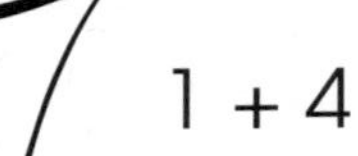
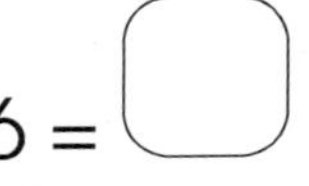
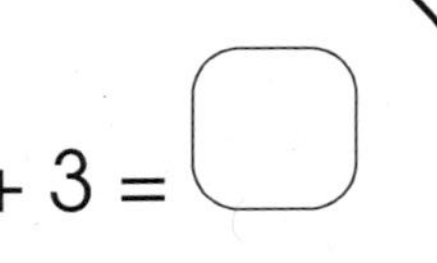
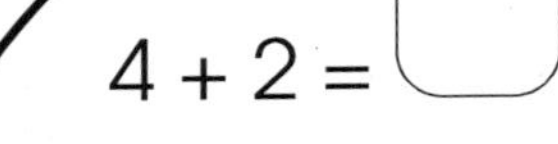

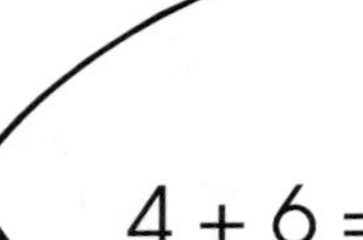
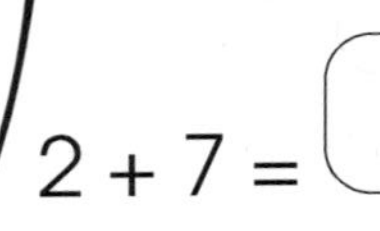
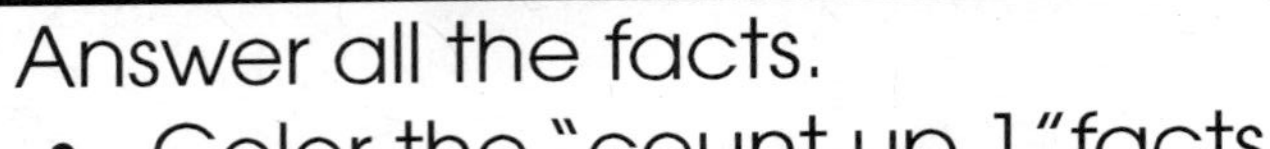
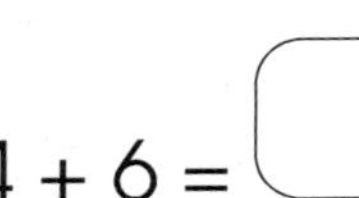
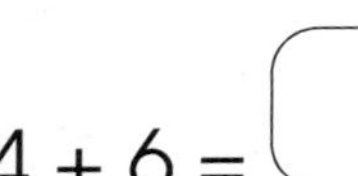
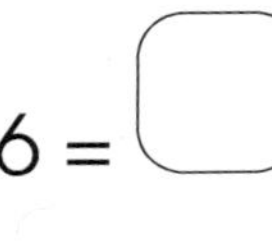
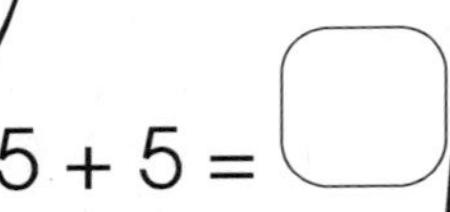
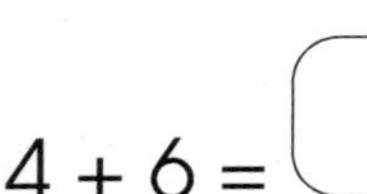

Answer all the facts.

- Color the "count up 1" facts blue.
- Color the "count up 2" facts green.
- Color the "count up 3" facts yellow.
- Color the "doubles" orange.
- Color the "magic 10" facts red.

7 + 3 = ☐

4 + 4 = ☐

6 + 1 = ☐

6 + 3 = ☐

2 + 7 = ☐

4 + 3 = ☐

4 + 2 = ☐

6 + 4 = ☐

2 + 2 = ☐

3 + 3 = ☐

4 + 6 = ☐

5 + 5 = ☐

3 + 6 = ☐

1 + 4 = ☐

3 + 2 = ☐

7 + 1 = ☐

All the facts to 10

Speed Test

① Answer the fast facts.

(a) Count up 1 from these numbers.

(b) Count up 2 from these numbers.

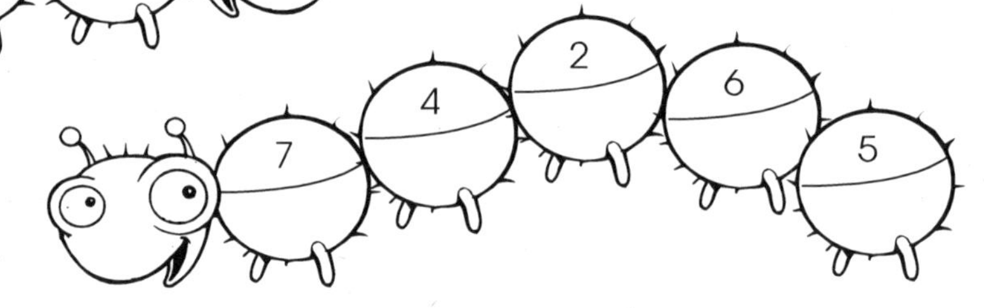

(c) Count up 3 from these numbers.

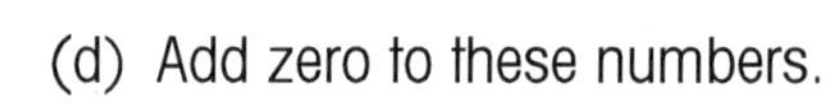

(d) Add zero to these numbers.

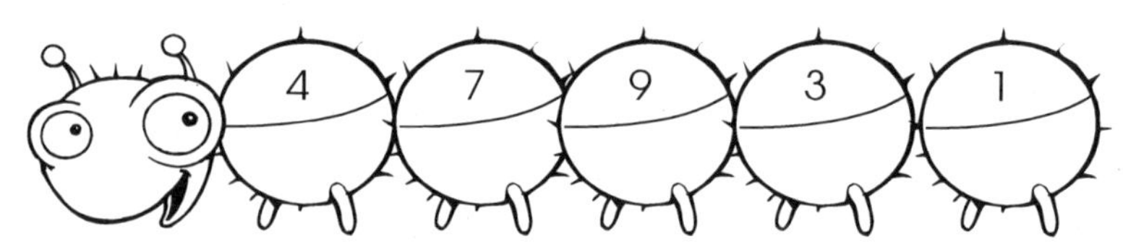

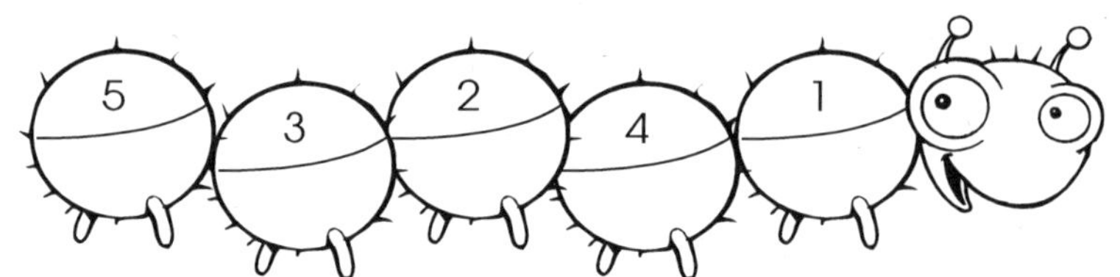

(e) Double these numbers.

② Answer these facts to 10.

(a) $1 + 7 =$ ☐

(b) $3 + 3 =$ ☐

(c) $9 + 0 =$ ☐

(d) $4 + 4 =$ ☐

(e) $1 + 9 =$ ☐

(f) $3 + 6 =$ ☐

(g) $5 + 5 =$ ☐

(h) $2 + 5 =$ ☐

(i) $8 + 0 =$ ☐

(j) $2 + 8 =$ ☐

(k) $3 + 4 =$ ☐

(l) $2 + 7 =$ ☐

(m) $6 + 1 =$ ☐

(n) $1 + 1 =$ ☐

(o) $0 + 5 =$ ☐

All the facts to 20

① Count up 1.

(a) 11 + 1 = ☐ (b) 16 + 1 = ☐ (c) 17 + 1 = ☐

(d) 1 + 19 = ☐ (e) 1 + 10 = ☐

How did you do?

1	2	3	4	5

② Count up 2.

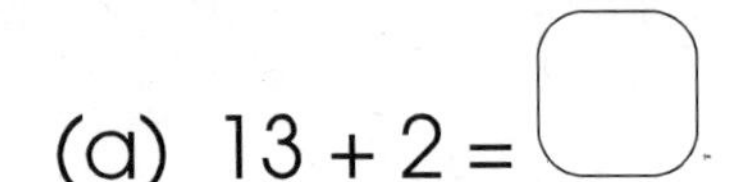

(a) 13 + 2 = ☐ (b) 2 + 15 = ☐ (c) 16 + 2 = ☐

(d) 17 + 2 = ☐ (e) 2 + 12 = ☐

How did you do?

1	2	3	4	5

③ Count up 3.

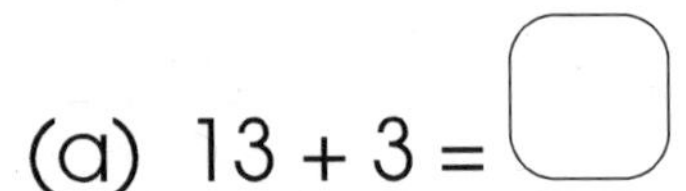

(a) 13 + 3 = ☐ (b) 17 + 3 = ☐ (c) 3 + 12 = ☐

(d) 3 + 11 = ☐ (e) 3 + 15 = ☐

How did you do?

1	2	3	4	5

④ Double.

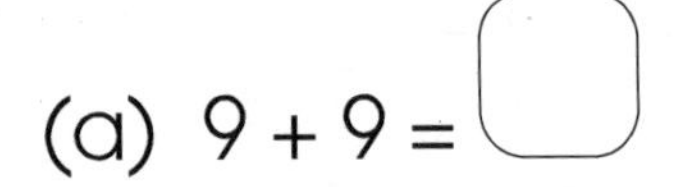

(a) 9 + 9 = ☐ (b) 7 + 7 = ☐ (c) 10 + 10 = ☐

(d) 6 + 6 = ☐ (e) 8 + 8 = ☐

How did you do?

1	2	3	4	5

⑤ Add 10.

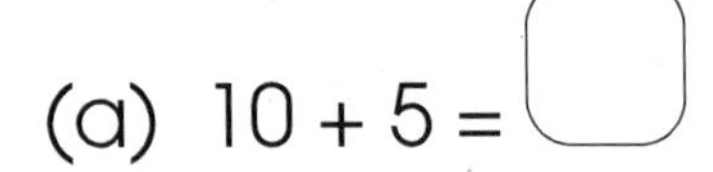

(a) 10 + 5 = ☐ (b) 10 + 8 = ☐ (c) 10 + 9 = ☐

(d) 2 + 10 = ☐ (e) 4 + 10 = ☐

How did you do?

1	2	3	4	5

All the facts to 20

Answer all the facts.

- Color the "count up 1" facts blue.
- Color the "count up 2" facts green.
- Color the "count up 3" facts yellow.
- Color the "doubles" orange.
- Color the "add 10" facts red.

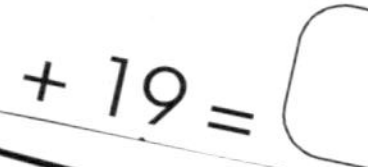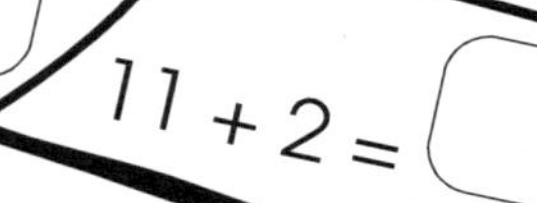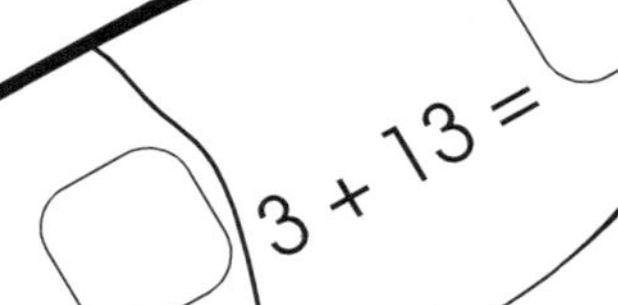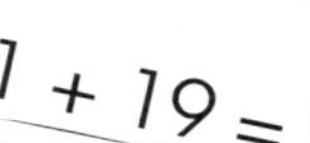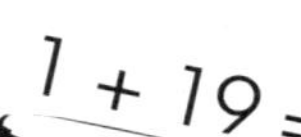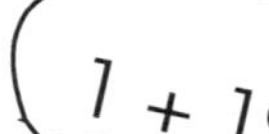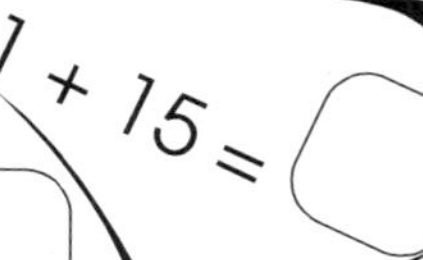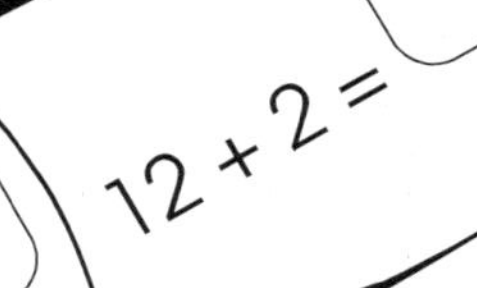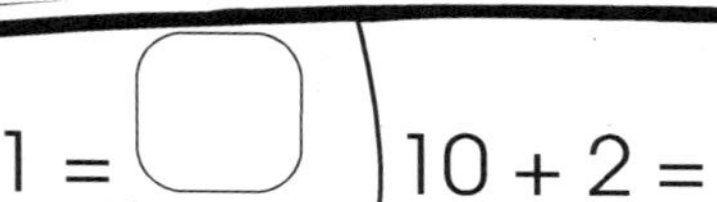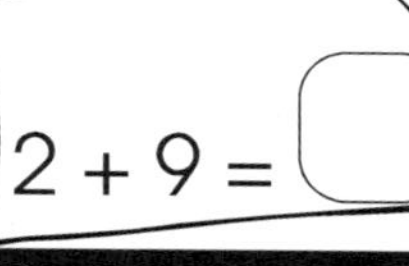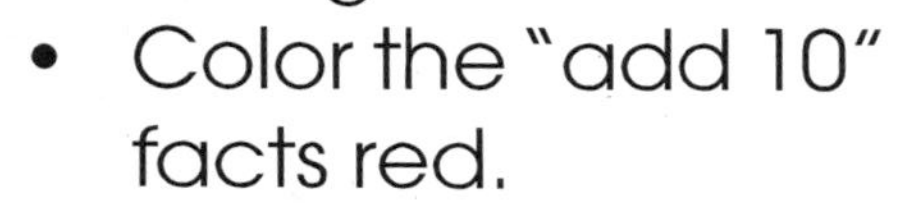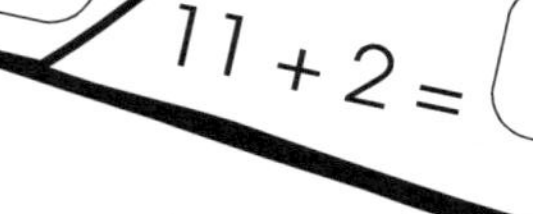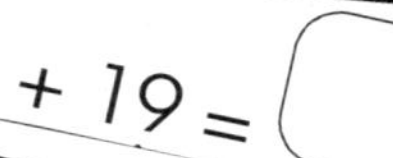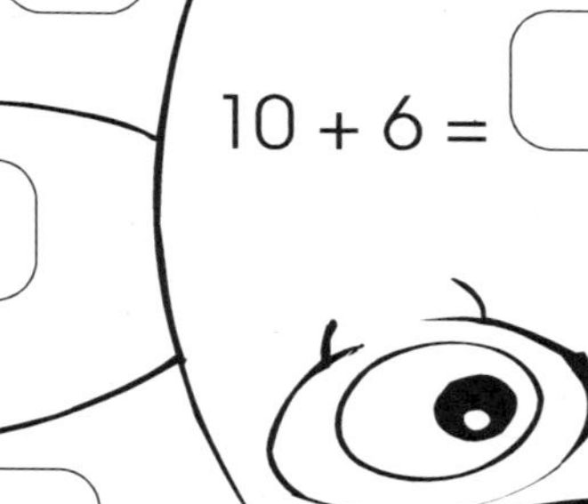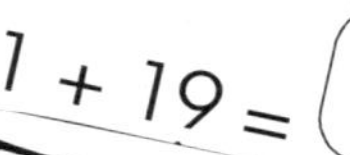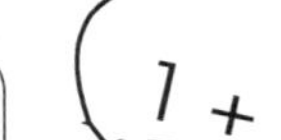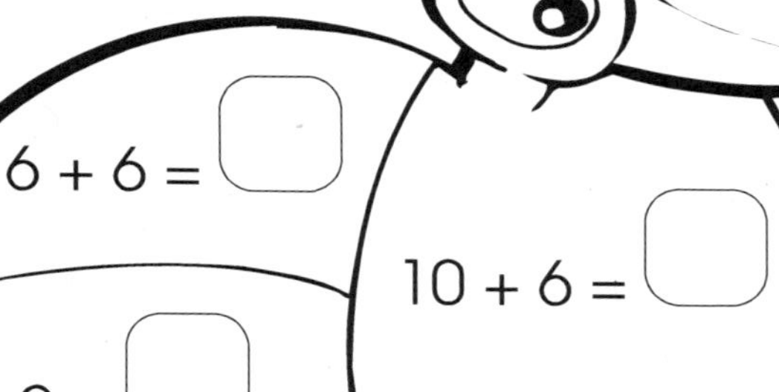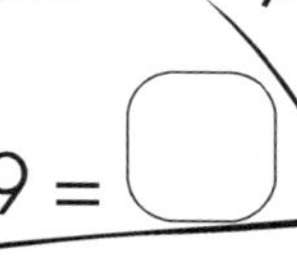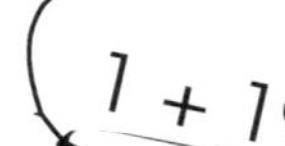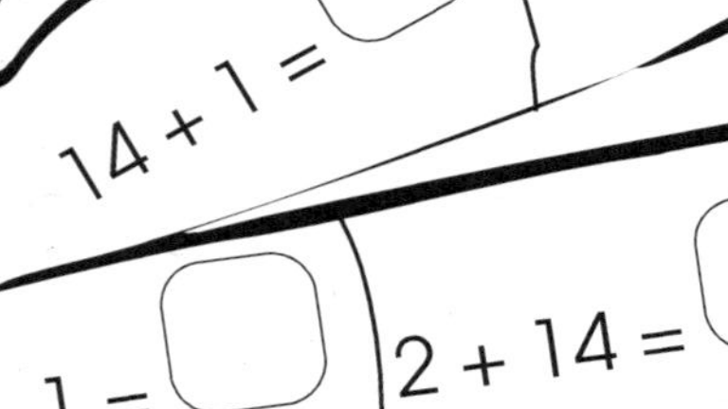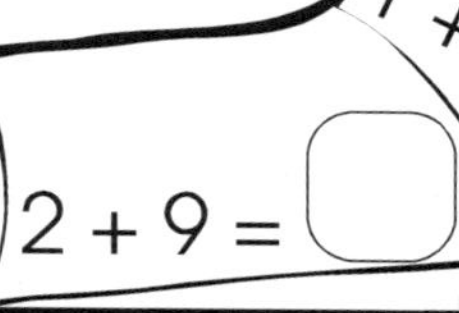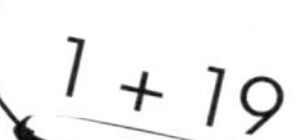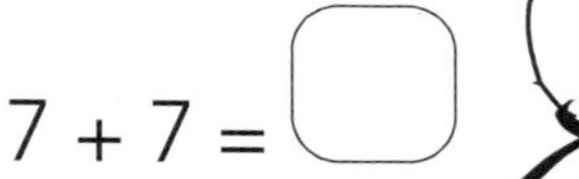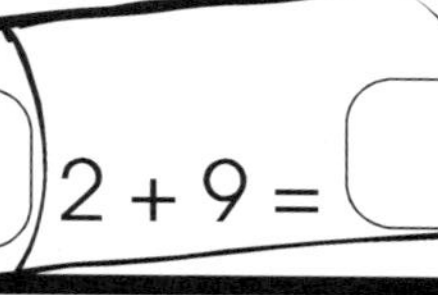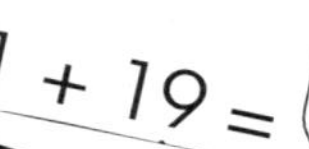

8 + 3 =

13 + 2 =

10 + 1 =

15 + 3 =

2 + 17 =

17 + 1 =

3 + 13 =

12 + 2 =

14 + 1 =

9 + 3 =

2 + 14 =

12 + 1 =

4 + 10 =

3 + 17 =

10 + 2 =

11 + 1 =

1 + 19 =

15 + 2 =

12 + 3 =

16 + 1 =

11 + 2 =

3 + 10 =

1 + 15 =

6 + 6 =

10 + 6 =

7 + 7 =

9 + 9 =

8 + 8 =

11 + 3 =

2 + 9 =

All the facts to 20
Speed Test

① Answer the fast facts

(a) Add 10 to these numbers.

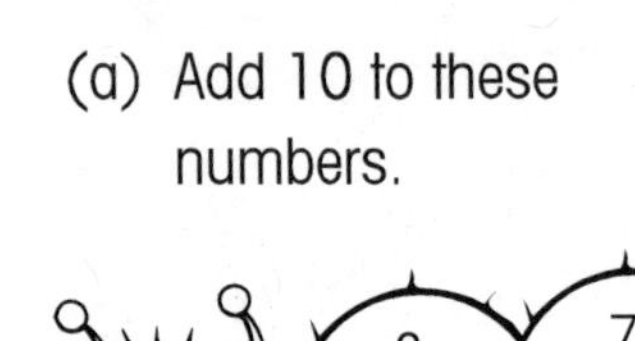
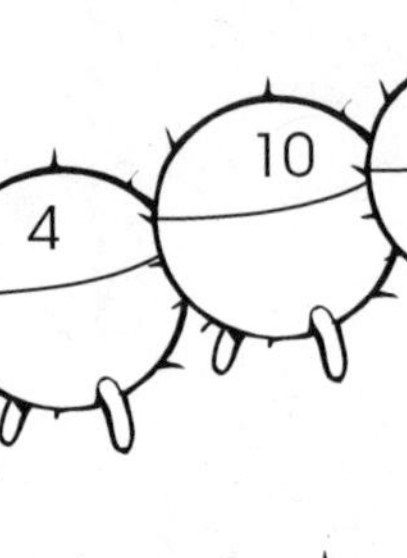

(b) Count up 2 from these numbers.

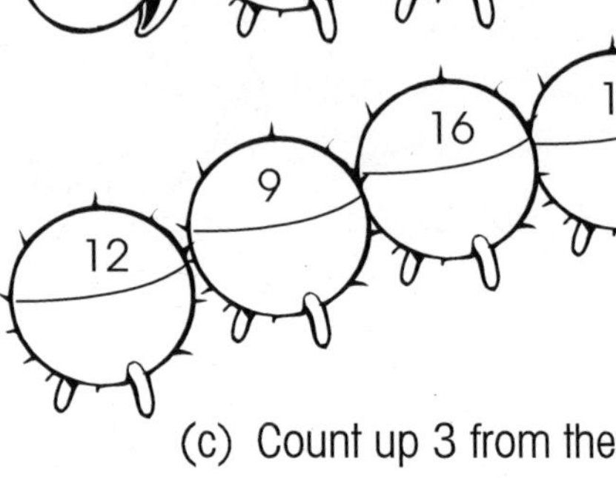

(c) Count up 3 from these numbers.

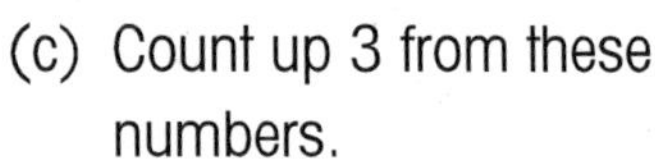

(d) Count up 1 from these numbers.

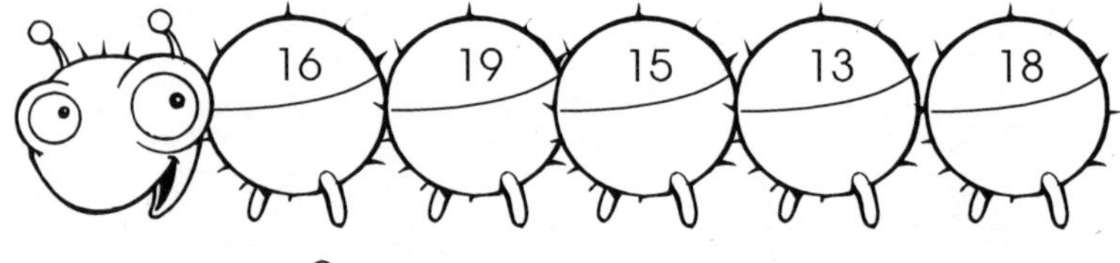

(e) Double these numbers.

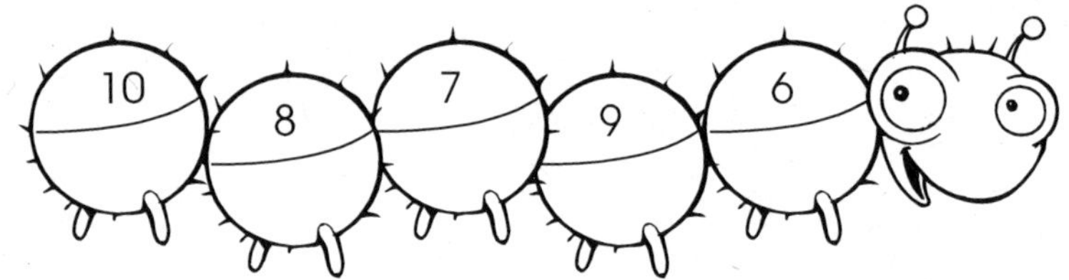

② Build these facts to 10 in your head to help you answer the number sentences.

(a) 7 + 4 = ☐ (b) 7 + 9 = ☐ (c) 5 + 6 = ☐

(d) 6 + 8 = ☐ (e) 5 + 8 = ☐ (f) 7 + 8 = ☐

(g) 4 + 8 = ☐ (h) 8 + 9 = ☐ (i) 9 + 5 = ☐

(j) 7 + 6 = ☐ (k) 9 + 6 = ☐ (l) 5 + 7 = ☐

Assessment
Strategies for adding to 10

Name: ______________________ Date: ________________

① Finish the number story. Draw pictures to help you.

3 jelly beans + 4 gummy worms = ☐ candies

② Circle and answer the "count up 1" problems.

(a) 2 + 5 = ☐ (b) 3 + 1 = ☐ (c) 1 + 6 = ☐

③ Circle and answer the "count up 2" problems.

(a) 2 + 5 = ☐ (b) 3 + 1 = ☐ (c) 7 + 2 = ☐

④ Circle and answer the "count up 3" problems.

(a) 6 + 2 = ☐ (b) 4 + 3 = ☐ (c) 3 + 6 = ☐

⑤ Answer these "zero" facts.

(a) 0 + 10 = ☐ (b) 0 + 1 = ☐ (c) 9 + 0 = ☐

⑥ Circle and answer the "magic 10" facts.

(a) 1 + 9 = ☐ (b) 7 + 3 = ☐ (c) 5 + 6 = ☐

Indicators

- Completes number stories to 10 successfully. 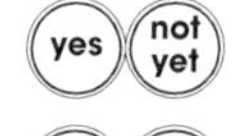

- Identifies and calculates "count up 1" facts. 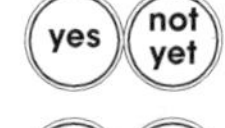

- Identifies and calculates "count up 2" facts. yes / not yet
- Identifies and calculates "count up 3" facts. 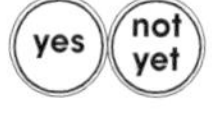

- Understands and adds zero correctly. 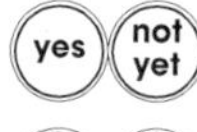

- Identifies the pairs of numbers that add to 10. 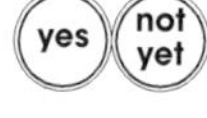

Calculates mentally | Requires concrete materials

Assessment
Strategies for adding to 20

Name: ______________________________ Date: ____________________

① Count up 1, 2, or 3.

(a) 2 + 11 = ☐ (b) 9 + 3 = ☐ (c) 1 + 12 = ☐

② Circle and answer the "double" facts.

(a) 7 + 7 = ☐ (b) 1 + 10 = ☐ (c) 8 + 8 = ☐

(d) 6 + 7 = ☐ (e) 9 + 9 = ☐ (f) 10 + 10 = ☐

③ Answer these "add 10" facts.

(a) 10 + 4 = ☐ (b) 6 + 10 = ☐ (c) 9 + 10 = ☐

④ Circle and answer the "double + 1" facts.

(a) 6 + 7 = ☐ (b) 5 + 7 = ☐ (c) 9 + 8 = ☐

(d) 5 + 5 = ☐ (e) 2 + 3 = ☐ (f) 7 + 7 = ☐

⑤ Build these facts to 10 to help you answer them.

(a) 7^{+}☐ + 8^{-}☐ = ☐ *becomes* 10 + ☐ = ☐

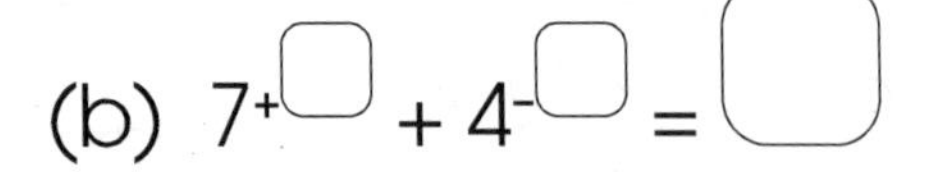

(b) 7^{+}☐ + 4^{-}☐ = ☐ *becomes* 10 + ☐ = ☐

Indicators

- Counts up 1, 2, or 3 to assist addition  yes / not yet
- Recognizes and answers "double" facts 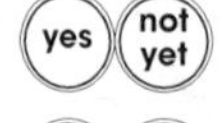 yes / not yet
- Recognizes and calculates "double + 1" facts  yes / not yet
- Adds 10 correctly.  yes / not yet
- Understands how to "build to 10" to help solve additions to 20. yes / not yet

Calculates mentally | Requires concrete materials

Answers

Number Stories for 2 .... page 6

1. Teacher check
2. Teacher check; 2

.... page 7

1. Teacher check
2. Teacher check

Number Stories for 3 .... page 8

1. Teacher check
2. Teacher check

.... page 9

1. Teacher check
2. Teacher check

Number Stories for 4 .... page 10

1. Teacher check
2. Teacher check

.... page 11

1. Teacher check
2. Teacher check

Number Stories for 5 .... page 12

1. Teacher check
2. Teacher check

.... page 13

1. Teacher check
2. Teacher check

Number Stories for 6 .... page 14

1. Teacher check
2. Teacher check

.... page 15

1. Teacher check
2. Teacher check

Number Stories for 7 .... page 16

1. Teacher check
2. Teacher check

.... page 17

1. Teacher check
2. Teacher check

Number Stories for 8 .... page 18

1. Teacher check
2. Teacher check

.... page 19

1. Teacher check
2. Teacher check

Number Stories for 9 .... page 20

1. Teacher check
2. Teacher check

.... page 21

1. Teacher check
2. Teacher check

Number Stories for 10 .... page 22

1. Teacher check
2. Teacher check

.... page 23

1. Teacher check
2. Teacher check

Assessment – Number Stories ... page 24

1. 2 fish and 3 whales make 5 sea creatures
2. 3 fish and 4 whales make 7 sea creatures
3. Teacher check

Count up 1 .... page 25

1. 2
2. 7
3. 10
4. 8
5. 4
6. 9

.... page 26

1. (a) 9 (b) 3 (c) 5;
   Teacher check
2. (a) 7 (b) 5 (c) 6 (d) 8;
   Teacher check

Count up 2 .... page 27

1. (a) 2, 3 (b) 5, 6 (c) 6, 7 (d) 9, 10
2. (a) 9 (b) 6 (c) 3 (d) 5;
   Teacher check

.... page 28

1. (a) 5 (b) 9 (c) 10 (d) 8
   (e) 7 (f) 5;
   Teacher check
2. (a) 8 (b) 10 (c) 8 (d) 9
   (e) 7 (f) 5;
   Teacher check
3. Teacher check

Count up 3 .... page 29

1. (a) 7, 8, 9; 9 (b) 8, 9, 10; 10
   (c) 6, 7, 8; 8
2. (a) 5 (b) 9 (c) 6;
   Teacher check

.... page 30

1. (a) 8 (b) 7 (c) 9 (d) 10;
   Teacher check
2. (a) 8 (b) 7 (c) 10 (d) 10
   (e) 8 (f) 5;
   Teacher check
3. Teacher check

Equal .... page 31

1. Teacher check
2. Teacher check
3. Teacher check
4. Teacher check

Adding Zero .... page 32

1. (a) 5 + 0 = 5
   (b) 3 + 0 = 3
2. (a) 0 (b) 0 (c) 4 (d) 0
   (e) 0 (f) 0
2. (a) 1 (b) 4 (c) 7

"Count up" Race .... page 33

1.	10	11.	8
2.	9	12.	10
3.	10	13.	7
4.	10	14.	7
5.	10	15.	2
6.	8	16.	10
7.	9	17.	8
8.	9	18.	6
9.	10	19.	7
10.	8	20.	7

You're Joking! .... page 34

A.	6	N.	13
B.	12	O.	14
C.	4	P.	14
D.	18	Q.	8
E.	17	R.	7
F.	5	S.	13
G.	30	T.	3
H.	19	U.	10
I.	11	V.	15
J.	19	W.	16
K.	10	X.	9
L.	18	Y.	19
M.	18	Z.	17

Answer: A fruit bat

Swap-around Facts (Count up 1) ... page 35

1. Teacher check; 10, 7, 7, 10, 3, 3
2. Teacher check; 5, 6, 6, 5, 9, 7, 7, 9
3. 4, 1 + 3 = 4;
   9, 8 + 1 = 9

Swap-around Facts (Count up 2) ... page 36

1. 10, 8, 8, 5, 5, 10
2. Teacher check; 6, 9, 9, 7, 6, 7, 10, 10
3. 6, 2 + 4 = 6;
   9, 2 + 7 = 9

Swap-around Facts (Count up 1) ... page 37

1. Teacher check; 7, 10, 9, 9, 10, 7
2. Teacher check; 8, 10, 7, 8, 10, 7, 9, 9
3. 9, 6 + 3 = 9;
   11, 8 + 3 = 11

"Swap-around" Street .... page 38

1. Teacher check
   (a) 3 (b) 5 (c) 7 (d) 6
   (e) 8 (f) 10 (g) 9 (h) 6
   (i) 8 (j) 10 (k) 9 (l) 9
   (m) 3 (n) 6 (o) 7 (p) 6
   (q) 9 (r) 8 (s) 9 (t) 5
   (u) 10 (v) 10 (w) 8 (x) 9
2. Teacher check
3. Teacher check

Doubles – 1, 2, 3 .... page 39

1. (a) 2 (b) 4 (c) 6
2. (a) Teacher check; 4
   (b) Teacher check; 6
   (c) Teacher check; 2

.... page 40

1. (a) 2 (b) 4 (c) 6;
   Teacher check

Doubles – 4, 5, 6 .... page 41

1. (a) 8 (b) 10 (c) 12
2. (a) Teacher check; 12
   (b) Teacher check; 8
   (c) Teacher check; 10

.... page 42

1. Teacher check
   (a) 8 (b) 10 (c) 12;

Answers

Doubles – 7, 8, 9 .... page 43

1. (a) 14 (b) 16 (c) 18
2. (a) Teacher check; 14
   (b) Teacher check; 18
   (c) Teacher check; 16

.. page 44

1. Teacher check
   (a) 14 (b) 16 (c) 18

Double Trouble! .... page 45

Teacher check

Double – Review .... page 46

1. (a) 2 (b) 4 (c) 6 (d) 8
   (e) 10 (f) 12 (g) 14 (h) 16
   (i) 18
2. (a) 6; 6, 12 (b) 9; 9, 18
   (c) 3; 3, 6 (d) 5; 5, 10
   (e) 8; 8, 16 (f) 4; 4, 8

.. page 47

6, 10, 7, 9, 18, 12, 10, 11, 16, 7, 11, 9, 10, 7, 8, 12, 14, 8, 7, 7

Double and Count up 1 .... page 48

1. 11; 5, 1
2. 15; 7, 1
3. 9; 4, 1
4. 13; 6, 1
5. 17; 8, 1

Doubles and Doubles Count up 1 .... page 49

1. 3	11. 3	21. 15
2. 11	12. 13	22. 11
3. 9	13. 13	23. 9
4. 17	14. 8	24. 13
5. 6	15. 18	25. 12
6. 17	16. 10	26. 3
7. 17	17. 11	27. 7
8. 13	18. 5	28. 15
9. 9	19. 11	29. 17
10. 5	20. 7	30. 3

"Magic" 10 Facts .... page 50

1. Teacher check
2. 4, (a) 2 (b) 7 (c) 5

.. page 51

1. (a) 9 (b) 7 (c) 6 (d) 4
   (e) 3 (f) 1 (g) 9 (h) 7
   (i) 6 (j) 4
2. Teacher check

Using "Magic 10" Facts .... page 52

1. (a) 13; 3, 13
   (b) 13; 3, 13
2. Teacher check
   (a) 13; 3, 13
   (b) 11; 1, 11
   (c) 11; 1, 11
   (d) 15; 5, 15
4. (a) 12 (b) 13 (c) 11

Building to 10 .... page 53

1. (a) 13; 3, 13
   (b) 15; 5, 15
   (c) 12; 2, 12
2. (a) 13; 3, 13
   (b) 11; 1, 11
   (c) 14; 4, 14
3. (a) 12; 2, 12
   (b) 15; 5, 15
   (c) 11; 1, 11

Adding on to 10 .... page 54

1. (a) 13 (b) 16 (c) 19 (d) 12
   (e) 17
2. (a) 11 (b) 14 (c) 18 (d) 13
3. Teacher check;
   14, 17, 16, 11, 19, 11, 16, 14, 19, 17

All the facts to 10 .... page 55

1. (a) 5 (b) 10 (c) 7 (d) 9
   (e) 8
2. (a) 5 (b) 8 (c) 6 (d) 9
   (e) 7
3. (a) 8 (b) 9 (c) 7 (d) 10
   (e) 8
4. (a) 10 (b) 10 (c) 10 (d) 4
   (e) 10
5. (a) 7 (b) 3 (c) 5 (d) 4
   (e) 10

.. page 56

Teacher check

.. page 57

1. (a) 7, 10, 6, 3, 8, 5, 2, 9, 4
   (b) 7, 8, 5 9, 6
   (c) 10, 7, 5, 9, 8
   (d) 4, 7, 9, 3, 1
   (e) 10, 6, 4, 8, 2
2. (a) 8 (b) 6 (c) 9 (d) 8
   (e) 10 (f) 9 (g) 10 (h) 7
   (i) 8 (j) 10 (k) 7 (l) 9
   (m) 7 (n) 2 (o) 5

All the facts to 20 .... page 58

1. (a) 12 (b) 17 (c) 18 (d) 20
   (e) 11
2. (a) 15 (b) 17 (c) 18 (d) 19
   (e) 14
3. (a) 16 (b) 20 (c) 15 (d) 14
   (e) 18
4. (a) 18 (b) 14 (c) 20 (d) 12
   (e) 16
5. (a) 15 (b) 18 (c) 19 (d) 12
   (e) 14

.. page 59

Teacher check

.. page 60

1. (a) 13, 17, 14, 20, 16, 12, 18, 15, 11, 19
   (b) 14, 11, 18, 16, 20
   (c) 14, 18, 11, 16, 20
   (d) 17, 20, 16, 14, 19
   (e) 20, 16, 14, 18, 12
2. (a) 11 (b) 16 (c) 11 (d) 14
   (e) 13 (f) 15 (g) 12 (h) 17
   (i) 14 (j) 13 (k) 15 (l) 12

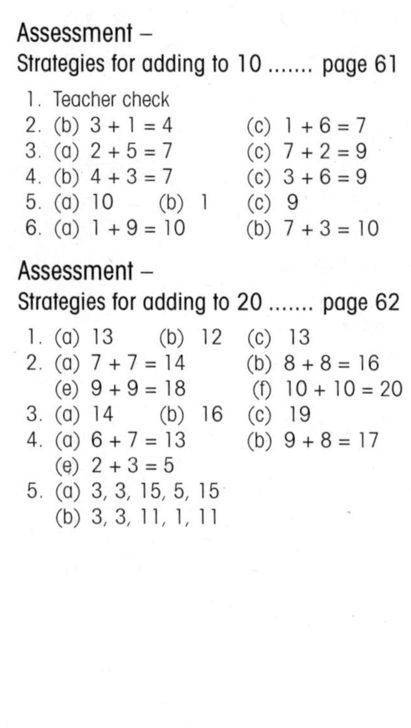

Assessment – Strategies for adding to 10 .... page 61

1. Teacher check
2. (b) 3 + 1 = 4 (c) 1 + 6 = 7
3. (a) 2 + 5 = 7 (c) 7 + 2 = 9
4. (b) 4 + 3 = 7 (c) 3 + 6 = 9
5. (a) 10 (b) 1 (c) 9
6. (a) 1 + 9 = 10 (b) 7 + 3 = 10

Assessment – Strategies for adding to 20 .... page 62

1. (a) 13 (b) 12 (c) 13
2. (a) 7 + 7 = 14 (b) 8 + 8 = 16
   (e) 9 + 9 = 18 (f) 10 + 10 = 20
3. (a) 14 (b) 16 (c) 19
4. (a) 6 + 7 = 13 (b) 9 + 8 = 17
   (e) 2 + 3 = 5
5. (a) 3, 3, 15, 5, 15
   (b) 3, 3, 11, 1, 11